Simulation

International Trends in Manufacturing Technology

SIMULATION

APPLICATIONS IN MANUFACTURING

Edited by
Dr. R.D. Hurrion

IFS (Publications) Ltd, UK

Springer-Verlag
Berlin   Heidelberg   New York   Tokyo
1986

British Library Cataloguing in Publication Data

Simulation: Applications in manufacturing.—(International trends in manufacturing technology)
    1. Industry — Simulation methods
    I. Hurrion, R.   II. Series
    338.4'767   HD38

ISBN 0-948507-33-0 IFS (Publications) Ltd
ISBN 3-540-16357-3 Springer-Verlag Berlin Heidelberg New York Tokyo
ISBN 0-387-16357-3 Springer-Verlag New York Heidelberg Berlin Tokyo

© 1986 **IFS (Publications) Ltd,** 35-39 High Street, Kempston,
               Bedford MK42 7BT, UK
  and **Springer-Verlag** Berlin Heidelberg New York Tokyo

Phototypeset by Wagstaffs Typeshuttle, Henlow, Bedfordshire
Printed and bound by Short Run Press Ltd, Exeter

International Trends in Manufacturing Technology

The advent of microprocessor controls and robotics is rapidly changing the face of manufacturing throughout the world. Large and small companies alike are adopting these new methods to improve the efficiency of their operations. Researchers are constantly probing to provide even more advanced technologies suitable for application to manufacturing. In response to these advances IFS (Publications) Ltd is publishing a series of books on topics that highlight the developments taking place in manufacturing technology. The series aims to be informative and educational.

Books already published in the series are:

Robot Vision, Programmable Assembly, Robot Safety, Robotic Assembly, Flexible Manufacturing Systems, Robot Sensors, Education and Training in Robotics, Robot Grippers and Human Factors.

Forthcoming titles include:

Robotic Welding, Automated Guided Vehicles, Artificial Intelligence, Electronics Assembly and Just-in-Time Manufacturing.

The series is intended for manufacturing managers, production engineers and those working on research into advance manufacturing methods. Each book is published in hard cover and is edited by a specialist in the particular field.

This, the tenth in the series – Simulation – is under the editorship of Dr. Robert Hurrion of the University of Warwick. The series editors are: Michael Innes and Brian Rooks.

Finally, the Publisher's gratitude is expressed to the authors whose works appear in this book.

Acknowledgements

IFS (Publications) Ltd wishes to express its acknowledgement and appreciation to the following publishers/organisations for granting permission to use some of the papers reprinted within this book, and for their cooperation and assistance throughout the production process.

IFS (Conferences) Ltd
35-39 High Street
Kempston
Bedford MK42 7BT
England

British Robot Association
28-30 High Street
Kempston
Bedford MK42 7AJ
England

Board of Directors
Winter Simulation Conference
PO Box 2413
West Lafayette
Indiana 47906
USA

British Aerospace plc
Business Development
  Department
Warton Division
Warton
Preston
England

Society of Manufacturing Engineers
One SME Drive
PO Box 930
Dearborn
Michigan 48121
USA

North-Holland Publishing Co.
PO Box 1991
1000 BZ-Amsterdam
The Netherlands

Systems International
Quadrant House
Sutton
Surrey SM2 5AS
England

Contents

1. Discrete-Event Simulation

Visual interactive modelling 3
R.D. Hurrion (University of Warwick, UK)

Animated CAD 15
L. Anderen (Cromemco, UK)

Applying simulation to assembly line design 21
K.J. Musselman and D.L. Martin (Pritsker & Associates Inc.,
USA) and J. Brouse (BDP Company, USA)

Computer simulation – A feasibility and planning tool for FMS 35
R.I. Mills (Ingersoll Engineers Inc., UK)

2. Risk Evaluation

Simulation and reduction of risk in financial decision-making 53
J.F. Wilson (Ingersoll Engineers Inc., UK)

Risk avoidance through independent simulation 61
P.L.C. Dunn (GEC Mechanical Handling Ltd, UK)

Simulation as a CIM planning tool 73
R.J. Miner (formerly of Pritsker & Associates Inc., USA)

3. Design of Machine and Robot Cells

Simulation of a robotic welding cell for small-batch production 89
R.W. Hawthorn, P.S. Monckton and R. Jones
(Wolverhampton Polytechnic, UK)

Modelling of a controller for a flexible manufacturing cell 105
T.S. Chan and H.A. Pak (University of Strathclyde, UK)

Manufacturing cell performance simulation using ECSL, 119
B.J. Clarke and P.F. Kelly (Huddersfield Polytechnic, UK)

4. Design of FMS

Simulation as an integral part of FMS planning 131
H.-J. Warnecke, R. Steinhilper and K.-P. Zeh (Fraunhofer-
Institut für Produktionstechnik und Automatisierung (IPA),
West Germany)

Decision support for planning flexible manufacturing systems 149
G. Seliger, B. Viehweger and B. Wienecke (Institut für
Produktionsanlangen und Konstruktionstechnik (IPK),
West Germany)

'FORSSIGHT' and its application to an FMS simulation study 163
M.J. Birch, T.J. Terrel and R.J. Simpson (Lancashire
Polytechnic, UK), and H.P. Feszczur (British Aerospace plc, UK)

Introducing FMS by simulation 173
A.S. Carrie and E. Adhami (University of Strathclyde, UK)

Experience in the use of computer simulation for FMS planning 185
T.C. Goodhead and T.M. Mahoney (University of Warwick
and Austin Rover Group, UK)

5. Control of Manufacturing Facilities

Designing the control of automated factories 199
R. Beadle (Istel Ltd, UK)

The use of simulation in cycle manufacturing 207
T.M. Gough (TI Raleigh Ltd, UK)

A production control aid for managers of manufacturing plants 213
P.W. Udo Graefe and A.W. Chan (National Research Council
of Canada) and M. Levi (Interfacing Technologies, Canada)

6. Simulation Standards

Simulation on microcomputers – The development of a visual 227
interactive modelling strategy
R.W. Hawkins, J.B. Macintosh and C.J. Shepherd
(Ford of Europe Inc., UK)

Computer simulation for FMS   243
N.R. Greenwood (General Electric Industrial Automation
Europe, UK), P. Rao (General Electric Corporate Research
and Development, USA), and M. Wisnom (Structural
Dynamics Research Corp., UK)

FMS: What happens when you don't simulate   251
J.E. Lenz (CMS Research Inc., USA)

Choosing and using a simulation system   261
R. Griffin (Inbucon Management Consultants Ltd, UK) and
A.H. Warby (IBM Ltd, UK)

Practical experience contrasting conventional modelling and   271
data-driven visual interactive simulation techniques
S.R. Hill and M.A.M. Rogers (ICI plc, UK)

7. Expert Simulation Environments

A knowledge-based system for simulation and control of FMS   287
D. Ben-Arieh (AT&T Bell Laboratories, USA)

AI-based simulation of advanced manufacturing systems   297
J. Shivnan (Digital Equipment International BV, Eire) and
J. Browne (University College Galway, Eire)

Expert systems and simulation in the design of an FMS   311
advisory system
B.R. Gaines (University of Calgary, Canada)

Authors' organisations and addresses   325

Source of material   329

Preface

This book describes the important role that simulation has played and will continue to play in manufacturing. The increased costs and complexity of manufacturing has made it vital for key manufacturing decisions to be thoroughly tested before implementation.

Computer simulation is a method by which models of alternative manufacturing scenarios may be developed and tested. The cost of changing and developing new manufacturing facilities is high so it has become essential to test out alternative manufacturing possibilities before implementation. This book describes the role that computer simulation can play in obtaining realistic models of proposed manufacturing configurations. Alternative manufacturing configurations and control strategies can be evaluated via computer simulation so that the best/robust/cost-effective solution may be implemented.

The book starts by introducing computer simulation and also describes the role that computer graphics now plays in the understanding of complex manufacturing models.

Computer simulation models used to be run on large mainframe computers. The results of any particular simulation run appearing in the form of summary print-out statistics. It is now common to link a computer simulation model with suitable animated computer graphics displays. Thus as the simulation runs an animated computer generated display is also produced. This method has allowed manufacturing management to check that the models are valid representations, and equally importantly have allowed manufacturing management to interact with the models in order to impart their experience.

The book then considers the general problem of risk evaluation, i.e. trying to improve the odds of a successful manufacturing enterprise. This is then followed by the role of simulation for the specific design of machine and robot cells.

Section 4 describes, via selected applications, the role of simulation in the design of flexible manufacturing systems. Section 5 shows how the simulation technique may be used to help plan and control existing manufacturing facilities. Since the simulation method is now becoming a routine management practice, Section 6 concentrates on simulation standards.

The techniques of computer simulation modelling is usually classified into two groups which are known as continuous and discrete-event simulation methods. Continuous simulation models tend to be used in order to model the physical characteristics of dynamic systems; for example, modelling the dynamics of an aircraft ejector seat, while discrete-event simulation models tend to concentrate on modelling the logical flow of elements, parts, customers through a system. The examples and applications described in this book are concerned with the logic flow and interaction of parts, machines, cells, AGVs ... for which the discrete-event simulation technique is more appropriate.

The final section looks forward to future developments of simulation with regard to manufacturing. The simulation technology is used by experienced managers, engineers and analysts. The emerging discipline of artificial intelligence, and in particular expert systems is considered. Can simulation models be assembled by expert programs as well as expert programmers? Is it possible for the expertise of managers and analysts to be incorporated in simulation models? The final section gives an introduction to this fascinating subject.

Bob Hurrion
September 1986

1

Discrete-Event Simulation –
An Introduction

The concept of simulation for most people would probably be associated with images such as aircraft pilots using simulators for training purposes. The role of simulation in manufacturing is in fact similar to this flight simulator analogy. Computer simulation models of existing or proposed manufacturing facilities can be developed so that management and engineers may 'fly' these models under both normal and adverse conditions. The learning associated with running manufacturing simulation models can then be fed back to give improved methods of control and design for manufacturing systems.

This section introduces simulation as a modelling technique and in particular describes the role that computer graphics now plays in helping to understand the interactions and dynamics of complex manufacturing systems.

VISUAL INTERACTIVE MODELLING

R.D. Hurrion
University of Warwick, UK

An approach which is becoming increasingly common for operational research applications is reviewed. The approach is known as 'visual interactive modelling'. It consists of first developing suitable visually animated graphics displays for the application under investigation. These displays supported by appropriate man–machine dialogues and models allow both analysts and users to explore dynamics of the application and thus gain an insight into its likely operational characteristics. The types of model, interactive dialogue and animated displays found in current visual interactive modelling applications are also reviewed.

During the past eight years the operational research community has seen a substantial increase in the use of computer graphics and microcomputing to support is traditional skills of mathematical model building. This paper reviews the emergence of an approach which has been of considerable assistance in numerous operational research applications. The approach is known generally as 'visual interactive modelling'.

The rationale of the approach is that an OR analyst builds a model (computer based) of some problem situation. The model has a graphics component so that the user commissioning the study can observe, in a suitable animated form, the dynamics of the model. The user is then able to use his or her own expert knowledge, experience and judgement of the original problem domain in order to interact with the model and experiment with alternative decision scenarios.

The key aspect of the visual interactive modelling approach, which differentiates it from other OR methods and techniques, is that it incorporates an animated graphics representation of the system under investigation. The improved communication, which the animated graphics representation brings to an OR study, is of particular importance at the multiple interfaces between client, analyst and model. The approach has helped managers gain confidence in the use of a model, they have then been better able to criticise it, while still contributing to its validation. In most visual modelling applications the modelling technique has become transparent, it is no longer regarded as a 'black-box' which in turn has helped the user gain a better insight into the original problem.

The idea of animating and allowing (OR) models to be interactive has taken some time to become established in the OR community. Donovan[1], Haverty[2], Bell[3], and Katske[4], describe early graphical facilities for simulation. Sulonen[5] documents the use of graphics to build block GPSS simulation models which are then run in a conventional batch manner and Sohnle[6] suggests general requirements for interactive simulation. Bazjanac[7] describes an animated interactive simulation which considers the emergency evacuation of people from high rise buildings using elevators, and Palme[8] comments on the advantage of having 'moving displays' generated by SIMULA simulation models. Hurrion[9] developed interactive simulation and animation methods for operational and production related problems. This lead to applications of visual interactive modelling (simulation) in batch manufacturing industries[10], process chemical plants[11-13], and facilities design in the automotive industry[14].

The majority of the early visual interactive modelling applications used simulation as its modelling mechanism, however Lembersky and Chi[15] developed an application technique which they describe as a decision simulator. In one of their examples a dynamic programming formulation is used to develop optimal forestry and tree cutting strategies. An interactive and animated graphics component was added to the model which then gave a realistic training environment for forestry staff.

The substantial cost reduction associated with interactive computing and the emergence of the microcomputer has meant that it is now possible to have relatively cheap but powerful local computing while access to mainframe resources such as large databases has also substantially improved. The combination of relatively cheap computer graphics terminals computing has meant that the idea of having animated interactive colour graphics to support the traditional OR modelling skills proposed in the '70's[16-18] is now both practicable and routine.

The software used by operational research practitioners, to support visual interactive modelling applications, is also becoming extensive; it can vary from small microcomputer menu-driven (visual interactive) graphics and modelling software through to the use of high-resolution graphics supported by mainframe computing. Pope[19,20] and Bell[21] catalogue current graphics animation for manufacturing systems and general visual modelling software. A conference, 'Simulation in Manufacturing',[22] also documents the role of graphics and animation for flexible manufacturing systems design.

This paper reviews visual interactive modelling which is the general term being used to encompass the interactive OR modelling skills which incorporate animation, visually dynamic and graphics representation.

Visual interactive modelling

The technique of visual interactive modelling consists of:

- Developing a model of the system under investigation.
- Incorporating a method of animating the model. This is usually achieved by showing the dynamics of the model on one or more colour graphics terminals. Having a dynamic animated view of the model ensures that the client commissioning the original study can observe the model in the form of a 'video' film. This has been shown to be of considerable help in removing the communications barrier that may exist between analyst and end-user with regard to the modelling technique and its assumptions. If a client or end-user can observe how a model is progressing through time then he/she is in a position to:
- Interact with the model in order to explore alternative decision strategies.

The ability of having a visual dynamic representation of a model; and the ability to interact with it are the key components of the visual interactive modelling approach.

The major contribution that visual interactive modelling has had, and probably will continue to have, is its ability to improve the communications and language barriers which exist between different management and professional staff for an application.

Visual interactive modelling is like a voyage of discovery[14]. The client/clients have their own perception of the target system. If the OR approach is perceived to be applicable, then the analyst/analysts will be employed to first develop a formal model. The analyst via interview and observations then establishes his own perceptions of the problem (quite often this will be in a mathematical form). The visual interactive model has tended to act as an interpreter between these different cultures. The technique thus allows the client to explore the current problem aided by the OR mathematical tools, but now having a window by which the client can understand them.

The main thrust of the visual modelling work has been to try to improve the communications between analyst and user so that both have an improved understanding of the problem. This improved understanding has lead to insights and has been the catalyst by which new policy rules have been articulated[14].

Visual modelling applications

Some visual interactive modelling applications are now described. The whole essence of the VIM approach is that a visual interactive dynamic model gives a better insight into an application than ever a static report will achieve.

The author is reminded of the school teacher's adage " a picture is worth a thousand words ...". With regard to visual interactive modelling the author thinks this should be extended to "... a colour video sequence is worth 1000 pictures while an animated interactive model is worth 1000 video sequences ...".

An example of the visual interactive modelling technique acting as a communications catalyst between an OR team and a production manager is documented by Brown[10]. This application is briefly described below.

... An OR project team had developed a complex manufacturing simulation which considered alternative batch scheduling rules. The two rules of particular interest were 'least stock per operation' (LSPO) and 'expect lateness' (EL).

The rules, when used in the simulation model gave answers which were opposite to what one might intuitively expect. The model and rules were believed to be correct but the OR team had substantial difficulty in convincing the production management staff as to their validity. The project stalled at this phase and any thoughts of implementation were not considered until the (simulation) model could be demonstrated as valid.

A simplified visual modelling experiment was developed in order to help communicate and understand the dynamics of various scheduling policies. The simplified model was:

Components enter a flow shop at a fixed rate of one item every six minutes. The flow shop consists of six machines. Each component must be processed in sequence at each machine, i.e. machine 1, followed by machine 2 ... concluding with machine 6. A component takes one minute to be processed at each machine and the transfer time between machines is assumed to be zero. An operator is also needed to assist in the processing of a component at each of the machines.

Suppose the first component arrives at time zero and that there is only one operator, then the question posed, once a suitable visual interactive model has been constructed, was:

If the work-in-progress at each machine consists of two components, what differences, if any, would occur if the rule 'process the next batch with the least number of operations remaining' was used or 'process the next batch with the most number of operations remaining' was used. Using the notation of Panwalker and Iskander[23], these priorities are FORNR (select the job that has the fewest operations) and MORNR (select the job that has the most operations remaining) respectively. The example problem is deterministic. Components arrive every six minutes and each component takes exactly six minutes of processing time.

This question was given to OR colleagues and production managers.

Participants were given only a few minutes in order to state their intuitive answer. The answers ranged from, no difference in work-in-progress between the rules, to some increase in work-in-progress if the rule 'most operations remaining' is used. (The reader is encouraged to answer this question before continuing, preferably by developing a visual model!)

Given the initial starting conditions then:

- Using the FORNR rule gives a stable solution of seven components waiting at machine 1 with one further component being progressed.
- Using the MORNR rule gives a stable solution of 43 components waiting for their final operation with one component being progressed.

This substantial increase in work-in-progress (over 500%) was not anticipated.

Clearly a complete prior analysis could have established the correct result; but in how many practical situations, because of the time demands on personnel would a complete analysis be undertaken? Production staff would prefer to trust to their own judgement. This simple visual interactive model was used as a catalyst which helped to unlock one fixed management perception.

There is a view that a major difficulty in management science and operational research practice is that of persuading the owner of a problem to change his mind[24]. If the manager and management scientist come to a problem with no preconceived notions then this task should be easier. Normally however, implementation of a project will involve the owner of the problem 'unmaking' his mind. The visual interactive modelling approach tends not to feed the manager with results but allows him to search for alternative solutions himself.

A further example showing the importance of animation in a training role is given by Lembersky and Chi[15]. They describe a forestry application which involves the optimal cutting of high-value North American West Coast Douglas Firs.

Operational decisions with regard to optimal tree cutting and logging have to be taken on site. A visual interactive model was developed, called a decision simulator, using high-resolution graphics to display trees. An operator, using a joystick, can rotate, turn, or generally view a tree from any angle. The operator can then simulate the cutting of the tree and then immediately compare his answers with the optimal solution available from a dynamic programming formulation. The authors report on the power of the visual model in this training environment. Forestry operators, with considerable practical experience were exposed to this 'decision simulator' which improved the quality of their operational tree and logging decisions. Savings of $7 million per year are reported.

Both of these examples were described in some detail to try to indicate to the reader the power, immediacy and learning which can occur from a visual interactive modelling session. In both examples the formal mathematical modelling techniques, i.e. comparison of scheduling rules or a dynamic programming algorithm, are known to the operational research practitioners. However the major difficulty of most OR studies is the transference of these modelling insights to the client. Numerous manufacturing and flexible manufacturing systems' visual modelling applications have been documented[19,20,25-30] (see also page 213). British Leyland used the technique for helping in the design of its Mini Metro plant and subsequent FMS applications[31].

One design problem considered the storage of car bodies between the spot welding and finish welding lines. After a car body has been spot welded it proceeds to a lift point where a sling transports it to a buffer store. The buffer store consists of a number of parallel conveyors. After leaving the store the sling and car body move to a drop section where a finish welding operation occurs. The plant layout also contained bypass loops so that slings can recycle. The original software which controlled the plant was such that on occasions lanes and conveyors became blocked.

A visual interactive simulation model of the store and its proposed control rules was built. Management and analysts were then able to observe an animated view of the plant. Running the model, observing its dynamic characteristics and evaluating alternatives were important stimuli in designing the control rules which now currently control the plant[14].

Further examples of animating the output of a flexible manufacturing system are described by Wortman and Wilson[32]. The application they describe uses colour graphics to model 10 horizontal milling machines, which are in turn serviced by a combination of raw material input, conveyor transportation, inspection and robot loading stations. The authors state in conclusion that "... a facility diagram can be animated to allow the engineers to actually see the operations ... such animations are key elements in the presentation of results to decision makers ..."

Nenonen[27,28] has developed visual interactive models with special reference to the mining industry. While Rogers[33] describes a visual model appropriate to the chemical industry.

The list of visual interactive modelling applications is substantial and growing. This section has included a sample of applications which was intended to give the reader an insight into the power that animated dynamic visual models can have in decision making.

Discussion

The actions of a visual interactive modelling application thus consist of developing a model, having the ability of dynamically animating the progress of it while giving the user the ability of interacting with it. These components are now considered separately.

A visual dynamic representation

The technique (VIM) is particularly concerned with obtaining a dynamic visual representation of the model. A key component in this regard is the communications or increased bandwidth that can flow from model to user if animated colour graphics techniques are used. It cannot be over stressed that the success of the visual interactive modelling technique appears to be mainly due to the fact that the end-user is in a far better position to understand the dynamics and assumptions of the model, if it is in a visually animated form. Applications to date have concentrated on two major types of visual display.

Schematic. The majority of visual interactive simulation applications have tended to use dynamic displays which try to mimic the operational characteristics of the system under study. This is particularly true of the manufacturing/flexible manufacturing systems/chemical applications where the development of suitable visual mimic or blueprint diagrams would seem to be natural method of displaying the model[11,31,34] (see also page 185.

The interactive and gaming simulations developed at British Steel also concentrated on schematic displays. Hollocks[35] documents the role of interactive simulation within British Steel. He describes interactive simulations which used physical or analogue display panels linked to the host simulation computer. He also describes the use of colour graphics terminals to display more recent simulation applications.

Logical. The second type of animated displays have tended to take the form of logical representations of the system under investigation. These logical representations have taken the form of bar charts, time series and histogram graphics. They have been used when the main component of the model is financial[21], or for complex models where it is sensible only to show summary measures of performance.

Interactive models

If a model has been animated then it becomes necessary to give the user the necessary feedback ability so that he/she can interactively change the model and its decisions. This interaction could be at the model development stage or while the model is running. Interaction at run-time was originally seen to be more important which lead to applications which used a compiled language[17]. The analyst then had to develop flexible model structures which were then compiled which allowed the user to make interactive parameter changes in the model. While this approach gave emphasis to interaction at run-time the analyst, when developing a model, had to use all his ingenuity to retain flexibility. For interactive simulation applications it became attractive to use the three-phase method[36]. This method breaks a complex simulation model down to a collection of activities which can be viewed as independent and hence relatively simple to modify and paramet-rise. The visual interactive modelling technique, using a compiled approach has been successful, it has undoubtedly influenced decision makers and numerous articles have cited anecdotal evidence of savings and improvements to decision processes.

However, this very success of making the model transparent to the client has shown the need to be able to interact with a model structure so that the analyst and client can both develop new components or decision rules to a model 'on the fly'. Withers[37] has commented on the fact that the interactive nature of simulation methods meant that they were limited to parameter changes at run-time. It is important during the use of a model for the client, analyst and model to act as equal partners in order to explore some insight. This would most certainly involve respecifying the model. For applications in a limited domain it is possible to specify the components needed in order to develop a model and thus provide the user with a menu driven interactive system which can build models in that limited domain. Withers[37] describes this principle for the domain of continuous flows of fluids between vessels. Norman[29] describes the approach for flexible manufacturing systems which use a limited set of components and O'Keefe[16], suggests using a menu driven interpreter approach which allows a model specification to be interactively modified at run-time.

One of the current difficulties of VIM is in fact due to the transparency that visual models give. An analyst will try to develop a robust, flexible model by the comprehensive use of parameters. It is difficult for users to know if their suggestions for model refinements can be easily accommodated (via parameter changes) or need substantial model recoding. The ability of extending the interactive nature of current VIM from run-time to include the model specification phase is a key issue[16,37].

The model or models

The third component to VIM is the model or models themselves. The majority of applications have used the simulation technique. However, Lembersky and Chi[15] describe their decision simulator approach for which one of their applications used a DP model. Da Silva[38] describes a complex batch scheduling problem which uses MRP. The results are displayed as a hierarchy of graphs and time series. The course plan can then be refined by a series of deterministic sub-models which consider only localised aspects of the MRP while Lapalme[39] and Fisher[40] describe interactive animated colour graphics interfaces for vehicle routing and truck dispatching applications.

Concluding remarks

Visual interactive modelling is becoming a routine practice for the management scientist. The relative cost of computer graphics facilities is continually reducing while the relative power of interactive computing is increasing.

In the near future it will be cost effective to supplement animated displays with video disk output. Interactive facilities also continue to be more cost effective, i.e. voice and tablet input are becoming convenient methods of interacting with models.

The emerging technology of expert systems (as an example of model) will also have animated graphics components. One of the key aspects of expert systems is their ability to explain their reasoning. An animated visual model would be an ideal method for an expert system to communicate with their users.

Visual interactive modelling techniques have improved the communication, understanding and insight of an OR model. It would seem likely that the fundamental concept of an OR model will be replaced by a visual interactive OR model.

References

[1] Donovan, J.J., Jones, M.M. and Alsop, J.W. 1968. A graphical facility for an interactive simulation system. In, *Proc. IFIP Congress 1968.* North-Holland, Amsterdam.

[2] Haverty, J.P. 1968. *Grail GPSS: Graphic On-Line Modelling.* Rand Corp., Santa Monica, CA, USA.

[3] Bell, T.E. 1969. *Computer Graphics for Simulation Problem-Solving.* Rand Corp., Santa Monica, CA, USA.

[4] Katske, J. 1975. User's Guide NGPSS: Superset of GPSS V for batch and interactive version. Norden Report 4059, R0001, Norwalk, CT, USA.

[5] Sulonen, R.K. 1972. On-line simulation with computer graphics. *On-Line*, 72.

[6] Sohnle, R.C., Tartar, J. and Sampson, J.R. 1973. Requirements for interactive simulation systems. *Simulation*, 20(5): 145-152.

[7] Bazjanac, V. 1976. Interactive simulation of building evacuation with elevators. In, *9th Annual Simulation Symp.*, Florida, March 1976, pp. 15-29.

[8] Palme, J. 1977. Moving pictures show simulation to user. *Simulation* 29(6): 240-209.

[9] Hurrion, R.D. 1976. The design use and required facilities of an interactive visual computer simulation language to explore production planning problems. PhD. Thesis, University of London.

[10] Brown, J.C. 1978. Visual interactive simulation: Further developments towards a generalised system and its use in three problem areas associated with a high-technology manufacturing company. MSc. Thesis, University of Warwick.

[11] Fisher, M.W.J. 1982. The application of visual interactive simulation in the management of continuous process chemical plants. PhD. Thesis, University of Warwick.

[12] Secker, R. 1977. That visual interactive simulation offers a viable technique for examining complex production planning and scheduling problems. MSc. Thesis, University of Warwick.

[13] Rubens, J. 1979. A study of the use of visual interactive simulation for decision making in a complex production system. MSc. Thesis, University of Warwick.

[14] Fiddy, E., Bright, J.G. and Hurrion, R.D. 1981. See-why: Interactive simulation on the screen. In, *Proc. Institute of Mechanical Engineers*, C293/81, pp. 167-172.

[15] Lomberoky, M.R. and Chi, U.II. 1904. Decision simulators speed implementation and improve operations. *Interfaces*, 14(4): 1-15.

[16] O'Keefe, R.M. 1984. Developing simulation models: An interpreter for visual interactive simulation. PhD. Thesis, University of Southmpton.

[17] Hurrion, R.D. and Secker, R.J.R. 1978. Visual interactive simulation, an aid to decision making. *Omega*, 6(5): 419-426.

[18] Alemparte, M., Chheda, D., Seeley, D. and Walker, W. 1975. Interacting with discrete simulation using on-line graphic animation. *Computers and Graphics*, 1: 309-318.

[19] Pope, D.N. 1984. A review of graphics animation of manufacturing systems. In, *AUTOFACT 6, Conf. Proc.* Anaheim, CA, October 1984.

[20] Pope, D.N. and Stroud, M.G. 1984. Animation of network-based simulation models. Presented at *ORSA/TIMS Joint National Meeting*, Dallas, November 1984.

[21] Bell, P.C. 1984. Benefits and challenges of decision simulator techniques. Presented at the *ORSA/TIMS Joint National Meeting*, Dallas, November 1984.

[22] Heginbotham, W.B. (Ed.) 1985. *Proc. 1st Int. Conf. on Simulation in Manufacturing*. IFS Publications, Bedford, UK, 1985.

[23] Panwalker, S.S. and Iskander, W. 1977. A survey of scheduling rules. *Operational Research Quarterly*, 25(1): 45-61.

[24] Boothroyd, H. 1978. *Articulate Intervention*. Taylor and Francis, London.

[25] Graefe, P.W. Udo 1982. Systems model of dragline for oil sands mining. Division of Mechanical Engineering. National Research Council of Canada.

[26] Looney, M.W. and Warby, A.H. 1985. Simulation in a high variety small batch FMS project – A case study. In, *Proc. 1st Int. Conf. on Simulation in Manufacturing*, pp. 367-376.

[27] Nenonen, L.K. 1982. Interactive computer modelling of truck haulage systems. *CIM Bulletin*, 75: 847.

[28] Nenonen, L.K., Graefe, P.W. Udo and Chan, A.W. 1981. Interactive computer modelling of truck/shovel operations in an open-pit mine. In, *Winter Simulation Conf. Proc.* pp. 133-139.

[29] Norman, T.A. 1984. Graphical simulation of flexible manufacturing systems. In, *Winter Simulation Conf. Proc.*

[30] Graefe, P.W. Udo and Chan, A.W. 1985. A production control aid for managers of manufacturing plants. In, *Proc. 1st Int. Conf. on Simulation in Manufacturing*, pp. 55-64.

[31] Anderen, L. 1984. Animated CAD. *Systems International*, November: 25-29.

[32] Wortman, D.B. and Wilson, J.R. 1984. Optimising a manufacturing plant by computer simulation. *Computer-Aided Engineering*, September.

[33] Rogers, M.A.M. 1978. Interactive computing as an aid to decision making. In, K.B. Haley (Ed.), *Operations Research '78*, pp. 829-842. North-Holland, Amsterdam.

[34] Goodhead, T.C. and Mahoney, T.M. 1985. Experience in the use of computer simulation for FMS planning. In, *Proc. 1st Int. Conf. on Simulation in Manufacturing*, pp. 229-238.

[35] Hollocks, B. 1983. Simulation and the micro. *Journal of the Operational Research Society*, 34: 331-343.

[36] Tocher, K.D. 1963. *The Art of Simulation*. English University Press, London.

[37] Withers, S.J. and Hurrion, R.D. 1982. The interactive development of visual simulation models. *Journal of the Operation Research Society*, 33: 973-976.

[38] Da Silva, C.M. 1982. The development of a DSS generator via action research. PhD. Thesis, University of Warwick.

[39] Lapalme, G., Cormier, M. and Mallette, C. 1983. A colour graphics system for interactive routing. Presented at the *ORSA/TIMS Joint National Meeting*, Orlando, November 1983.

[40] Fisher, M., Greenfield, A. and Thompson, K. 1983. Real world experience with an interactive colour interface for vehicle routing models. Presented at the *ORSA/TIMS Joint National Meeting*, Orlando, November 1983.

ANIMATED CAD

L. Anderen
Cromenco, UK

Modelling a factory or supermarket layout on a computer and simulating traffic flows can show up possible bottlenecks long before the first 'brick' has to be laid, let alone the placement of expensive machinery. This paper describes how micro-based, dynamic visual simulation helps the operational research specialist to define and tune operational procedures prior to implementation.

One of CAD/CAM's 'fringe' applications is the use of visual simulation techniques to assist the work of operational research (OR). And since it is an application which defies categorisation in any one specific area of computing, these highly advanced OR techniques rarely feature among the long list of CAD/CAM innovations.

Not surprisingly, one of the pioneers of visual simulation in OR is directly associated with a leading manufacturing company. In fact, it was British Leyland's system and software subsidiary, Istel which in its original livery of BL Systems launched the first dynamic, visual simulation package for the OR specialist.

Called 'SEE-WHY', one of its earliest applications was related to Austin Rover's Metro car plant at Longbridge. Here, the system proved itself as an OR tool which could, by simulating production flow, optimise a complex series of production and store processes, while minimising the cost and the size of the facilities needed to support production. For example, SEE-WHY was used to evaluate the feasibility of the control and the capacity of the body-in-white and painted-body stores, responsible respectively for releasing a controlled supply of bodies for painting and final assembly.

Simulating the overall layout of the factory, scheduling rules were identified to avoid bottlenecks in either of the stores and their conveyors, taking account of realistic rates of machine breakdown. A store too small would have seriously inhibited production flow, while one too large would have tied up far more capacity than was necessary.

Since then, Istel has successfully marketed its product to many UK and overseas companies involved in the manufacturing and process industries, as well as for other less classic simulation applications.

The system is based around a 16-bit microcomputer, together with a large, colour display to provide an animated representation of a complex simulation.

The system is, however, unusual in that it was first developed on an 8-bit microcomputer, few of which then had had enough performance or capacity to run large and complex models. Nonetheless, Istel succeeded in developing the system on a Cromemco 8-bit micro, the only one the company found that could run complex simulations fast enough and present pictures that moved at an even, realistic rate.

In the Metro project, the more traditional GPSS simulation package produced nothing which would help management to understand exactly what the proposals really meant in operational terms. It was first tried using a kind of board game, around which managers gathered to see what would be happening on the factory floor. Obviously, this was far too slow, so it was decided to computerise the process, so that one could show a whole day's production in minutes.

Variable response

The company started out using its IBM 3033 with the TSO option. However, the response was far too variable and much too slow when many people were using the system, with the net result that the simulation picture just stopped moving. They then scaled it down to a PDP-11/34 but, since the plan was to develop a package which could be sold elsewhere in BL and in external markets, this system would make the product too expensive – unless, of course, the user happened already to be using a PDP-11/34. So they went looking for a suitable micro.

Simulation requires a lot of pure CPU horsepower, which was very hard to find on an 8-bit micro. The Cromemco enabled the use of bank-switching. In other words, instead of multiple disk transfers, the program could actually switch between processor boards inside the machine. That enabled the necessary processing speed to make sure that the picture continued to move at an even rate.

SEE-WHY was later moved to the 16-bit Cromemco on which the current product now runs.

Visual dynamics

The OR specialist uses the system to construct and manipulate a complex model – mathematically, visually and interactively – in order to establish optimum control rules for running, say, process plant, batch and flow-line processes, or even the flow of passenger and baggage traffic through air terminals.

Control rules, for example, the rate at which items move through processes, how many, how they are routed according to their status or condition, and so on, are applied to the model and are tested for feasibility. As the simulation runs, a speeded-up picture shows the user where problems are likely to occur. The model can then be stocked, changed and played back at will to rectify problems. For instance, the user can increase capacity in certain areas, re-schedule or re-route the flow, and so on, until the model is actually *seen* to be viable.

The system is fully interactive, so that if the model makes a decision that the operator doesn't agree with, he can override it and see the effect. Most other simulation packages use more traditional techniques which record what has happened and then play back the picture at the end. It cannot be changed in the middle of the simulation.

To expand its use beyond the OR specialist, Istel has added a menu-driven user interface (Express) which will allow non-specialists to build a model interactively, with the system prompting and helping in the definition of the individual events that are to occur within the simulation.

Users thus no longer need to understand FORTRAN. Of course, users have to understand the basics of simulation so that they are able to structure the problem they are trying to solve, but most engineers are able to do this and so all they have to think about is the actual operation itself.

Examples

These examples, though very simply described, involve a vast range of complex variable and alternative strategies which have been tested, checked and changed to create viable, economic models which exactly represent live operational procedures. In most cases, users have made considerable savings on capital investment as a result of greater efficiency.

British Airports has used the system for a variety of applications: modelling aircraft movements and apron parking, for check-in and forecourt modelling plus looking at the effect of changes and mix in air traffic.

The first application involves finding the best method of scheduling the aircraft for landing and take-off. Some of the critical issues which have been considered include the amount of turbulence caused during take-off and landing of wide-bodied jets. Sufficient time has to be allowed before smaller aircraft take-off behind a Jumbo – the turbulence is enough to flip the smaller aircraft over. Therefore, air traffic control needs to send the wide-bodied jets down the runway one after the other, if it is to avoid delays in take-off procedures. From the 'stacking' aspect, it is a question of stacking aircraft so that all flights land at more or less their scheduled time, bearing in mind their fuel levels and the turbulence they cause when landing. Late flights have also to be considered, especially long-haul ones which must be landed in spite of the current schedule.

With the new Heathrow Terminal 4, British Airports used SEE-WHY to evaluate baggage handling and other facilities needed to handle the anticipated amount of traffic coming in and going out from the terminal. For example, how many baggage carousels are needed? How long should they be? How many immigration check and customs desks are needed? Are the facilities laid out such that it is easy for passengers to move through the terminal? All these and more issues were evaluated to establish what facilites were needed and they will be arranged for maximum operational convenience and speed.

The aircraft parking problem was evaluated. This involves trying to get as many passengers as possible to disembark at gates and proceed under cover to passport control, rather than to travel from the plane by bus. It may, for example, be better to park *all* frequent short-haul flights at the gates, and to park larger aircraft, which require more time to refuel and prepare for take-off, on the tarmac. British Airports now uses a SEE-WHY model to show its air traffic controllers how best to decide where incoming aircraft should be parked, and to which gates or areas they should be directed.

British Telecom is another user applying the system to a classic simulation problem of working out the best operational design for telephone exchanges–basically, how many lines are needed to service incoming calls, and how they should be serviced. Similarly, the British Post Office has used the system to establish the most efficient methods by which to sort and transport mail, using the system to develop its own specialised postal sorting model.

Refining

In Canada, the Exxon subsidiary Imperial Oil bought SEE-WHY to help schedule oil through its refining processes and has embedded it within its day-to-day control systems. Oil has to pass between a series of storage tanks according to its type, quality and suitability for mixing. The scheduler has to check which ships are to be unloaded and to decide where their cargo should to be sent. From the tanks, the oil is processed (cracked), mixed, turned into end-products and stores in further tanks ready for shipping.

The scheduler has to work out which tanks to use and which products to make in order to fully utilise process equipment. Mistakes in mixing could demand that further oil is flown in so that production schedules are met and crackers are kept working at their peak. Imperial is quoting that its increased efficiency could be saving the company around half a million dollars a day – the potential effect of a wrong decision.

Now, the scheduler examines the forward load of ships, decides to which tanks the oil will go and then simulates the production schedule as passed down from the order bank on the company's IBM mainframe. If potential problems are highlighted, he changes the tank and tries again until he finds the best solution, which is then passed back for the control systems to implement.

In the UK, Lucas has used its system to schedule the production of parts through a series of light engineering machine tools. For instance, the company wanted to produce a new product line using an existing machine shop already making a wide variety of electronic components. The engineers first estimated that in order to do this, the company would need to invest around £2.5 million in new equipment. However, when the OR specialists checked the increased loadings and scheduling via SEE-WHY, they managed to schedule the new products in without any fresh investment whatsoever.

Tills

Even banks and building societies could make major savings with this system where the investment in computerised tills is phenomenal. Banks involve relatively simple operations, but how do you decide whether it is better to have a series of tills serving several queues, or just one? Would specialised tills be an advantage, say, for foreign exchange, withdrawals and payments, or should they be multi-purpose? Are special banks needed? Building societies, now coming up to compete with banks, are investing heavily in computerised, on-line tills. Here, they could be saving thousands by optimising the tills and the computer needed to run their operations.

Automated warehousing and other materials handling facilities are other classic targets for visual simulation techniques. Here, control rules have to be established to minimise the number of cranes used and the distance they travel. Is it better to unload at the front and load from the back? Should the machine pick from each pallet in situ, or pull it out, pick and put it back again? Should one bulk pick or order pick? What sort of transport mechanisms are needed to move goods between stores and processes-conveyors, automated guided vehicles, or plain fork-lift trucks. Often, SEE-WHY will prove that cheaper, traditional methods are just as efficient for keeping production flow moving at acceptable rates, which can ultimately mean the difference between profit and loss on the investment in new facilities.

Of course, the system's most obvious application is in the car industry, where multi-million pound savings have been made on single factory processes. For example, saving on one body store can save something in the order of around £10 million in capital investment.

APPLYING SIMULATION TO ASSEMBLY LINE DESIGN

K.J. Musselman and D.L. Martin
Pritsker & Associates Inc., USA
and
J. Brouse
BDP Company, USA

The issues that must be addressed in designing assembly lines are numerous and varied. Because of this and the escalating cost of new equipment and lost production time, many companies, including Carrier, have begun to use simulation to help design these lines. Simulation modelling provides an excellent discipline for system design. It prompts for information and highlights relationships within the system that might otherwise not be considered important or even noticed until a much later date. It also gives the system designer a means of identifying and resolving problem areas in advance. The result is an improved design and increased confidence that the final system will operate as intended.

After years of planning, designing, engineering, justifying, and purchasing, the moment of truth has finally arrived – the installation of the new assembly line. And with its arrival comes instant anxiety: Will the new line achieve its designed throughput? Will there be enough buffer at the right locations? Will the controls maintain an even flow? Will more people and machines be required?

These questions are typical. But why must there be so much uncertainty now, when the line is being installed? Is the system so complex that a proper analysis cannot be done? Does testing of the line's control logic require an operational system? Do the cycle times of the machines and the assembly times of the workers vary so significantly that they preclude one from getting an accurate estimate of the system's throughput? Are the reliabilities of the machines so in doubt that proper compensation cannot be made when designing buffer locations and sizes?

These questions all touch on why it is so difficult to design an assembly line. Station cycle times vary, equipment jams, workers compensate, machines fall, and buffers fill. So how does one properly design an assembly line when the process itself behaves in such a dynamic and unpredictable way?

Simulation

One of the most effective ways to support the design of an assembly line, or any manufacturing system for that matter, is with computer simulation. By reproducing the activities of the system, a simulation model can provide insight into the system's overall behaviour and performance potential. It can address a wide range of questions, giving the design engineer an opportunity to be creative and innovative.

Examples of the types of questions that can be explored include:

- What happens if one of the test stations is removed?
- What happens if the size of the buffer in front of the repair station is reduced?
- Where should reworked parts enter the system?
- Should a unit be sent to the first available machine or to the machine that has been waiting the longest?
- What happens if the next unit at a staging point is held until the previous unit has cleared this segment of the conveyor?
- What must the reliability of a given station be to reach the system's overall throughput goal?
- What happens if machines are positioned across from each other, instead of in a line?
- Given the mix of products to be produced, how many pallets of each type are required?

Answers to questions such as these are possible through experimentation with a simulation model. Regardless of whether the system actually exists or not, the model can emulate the system's behaviour, helping to identify the real problem areas and determine the extent to which any proposed solution can solve these problems.

Simulation languages

Simulation is usually done using a language. The language is used to support the model building and analysis process. It does this by performing most of the functions basic to all simulation models, such as keeping system status information, advancing simulated time, ranking and selecting entities in the system, and collecting data.

There are two general classes of languages available: General Purpose and Special Purpose. General Purpose languages offer power and flexibility, for they are designed to address a wide range of issues. For instance, they can be used to determine the capacity of an urban storm drainage facility as well as the effect of a country's capital investment policies on the world. Examples of these languages include SLAM II[1], SIMSCRIPT II.5[2,3], and GPSS[4]. Special Purpose languages, on the other hand, offer more functionally specific

```
;
;      DEFINE SIMULATION (TITLE,DATE,NUMBER OF RUNS,LENGTH OF RUN)
;
BEGIN,P&A,ASSEMBLY LINE SIMULATION,4/15/1985,1,0,30600;
;
;      PART ROUTING (PART TYPE,ARRIVAL RATE,NEXT STATION,CYCLE TIME,
;                    CONDITIONS FOR ROUTING)
;
PART,PARTA,,8,0,1;
     ROUTE,WINDER,,8,,,TERMAPPL;
     ROUTE,TERMAPPL,,1,,,FOLD90,NPREPROCESS(FOLD90,PARTC).GE.1;
PART,PARTB,,3,0,1;
     ROUTE,FRAMEFAB,,3,,,ASSMB,NPREPROCESS(ASSMB,PARTD).GE.1;
PART,PARTC;
     ROUTE,REMPART,,0,,,UNFOLD90;
     ROUTE,UNFOLD90,,3,,,FOLD90;
PART,PARTD;
     ROUTE,FOLD90,,2,,,ASSMB;
PART,PARTE;
     ROUTE,ASSMB,,0,,,FOLD180;
     ROUTE,FOLD180,,3.5,,,INSULPL;
     ROUTE,INSULPL,,2,,,UNFLD180;
     ROUTE,UNFLD180,,3.5,,,REMPART;
     ROUTE,REMPART;
PART,PARTF;
     ROUTE,REMPART,,0,,,TEST;
     ROUTE,TEST,,2,,,GOOD,.99,BAD,.01;
     ROUTE,GOOD;
     ROUTE,BAD;
;
;      STATION DEFINITION (SIZE,BUFFER,TYPE)
;
STATION,WINDER,1,10,10,REGULAR/GANTRY1,,,,REGULAR;
STATION,TERMAPPL,1,,,REGULAR/GANTRY2,,,,REGULAR;
STATION,FOLD90,1,,,REGULAR/INTER,,,,ASSEMBLE,PARTD(PARTA,1,PARTC,1);
STATION,FRAMEFAB,1,10,,REGULAR/GANTRY3,,,,REGULAR;
STATION,FOLD180,1,,,REGULAR/TRANSFR2,,,,REGULAR;
STATION,INSULPL,1,,,REGULAR/TRANSFR3,,,,REGULAR;
STATION,UNFLD180,1,,,REGULAR/GANTRY4,,,,REGULAR;
STATION,UNFOLD90,1,,,REGULAR/BACK,,,,REGULAR;
STATION,TEST,1,,,REGULAR/GANTRY5,,,,REGULAR;
STATION,GOOD,1,,,;
STATION,BAD,1,,,;
STATION,ASSMB,1,,,REGULAR/TRANSFR1,,,,ASSEMBLE,PARTE(PARTD,1,PARTB,1);
STATION,REMPART,1,,,REGULAR/INTER,,,,PRODUCE,PARTE(PARTF,1,PARTC,1);
;
;      TRANSPORT DEVICE DEFINITION (NUMBER,TIME,SPEED)
;
TRANSPORTER,TRANSFR1,1,2;
TRANSPORTER,TRANSFR2,1,2;
TRANSPORTER,TRANSFR3,1,2;
TRANSPORTER,BACK,2,6;
TRANSPORTER,INTER,1,0;
TRANSPORTER,GANTRY1,1,4;
TRANSPORTER,GANTRY2,1,4;
TRANSPORTER,GANTRY3,1,4;
TRANSPORTER,GANTRY4,1,2;
TRANSPORTER,GANTRY5,1,,,,1;
DISTANCE,TEST,GOOD,2;
DISTANCE,TEST,BAD,4;
;
;      EQUIPMENT BREAKDOWNS (TIME BETWEEN,TIME TO REPAIR)
;
BREAKDOWN,TRANSPORTER,GANTRY1,EXPONENTIAL(2000),EXPONENTIAL(60);
BREAKDOWN,TRANSPORTER,GANTRY2,EXPONENTIAL(2000),EXPONENTIAL(60);
BREAKDOWN,TRANSPORTER,GANTRY3,EXPONENTIAL(2000),EXPONENTIAL(60);
BREAKDOWN,TRANSPORTER,GANTRY4,EXPONENTIAL(6000),EXPONENTIAL(60);
BREAKDOWN,TRANSPORTER,GANTRY5,EXPONENTIAL(6000),EXPONENTIAL(60);
BREAKDOWN,STATION,WINDER,EXPONENTIAL(6000),EXPONENTIAL(120);
BREAKDOWN,STATION,TERMAPPL,EXPONENTIAL(2000),EXPONENTIAL(60);
BREAKDOWN,STATION,FOLD90,EXPONENTIAL(1000),EXPONENTIAL(60);
BREAKDOWN,STATION,FRAMEFAB,EXPONENTIAL(6000),EXPONENTIAL(300);
BREAKDOWN,STATION,FOLD180,EXPONENTIAL(3500),EXPONENTIAL(60);
BREAKDOWN,STATION,INSULPL,EXPONENTIAL(4000),EXPONENTIAL(60);
BREAKDOWN,TRANSPORTER,TRANSFR1,EXPONENTIAL(4000),EXPONENTIAL(60);
BREAKDOWN,TRANSPORTER,TRANSFR2,EXPONENTIAL(4000),EXPONENTIAL(60);
BREAKDOWN,TRANSPORTER,TRANSFR3,EXPONENTIAL(4000),EXPONENTIAL(60);
BREAKDOWN,TRANSPORTER,BACK,EXPONENTIAL(12000),EXPONENTIAL(60);
;
;      INITIALIZE MODEL
;
ISTATION,PARTC,FOLD90,100;
ITRANSPORTER,PARTA,TERMAPPL,WINDER;
;
;      CLEAR STATISTICS
;
CLEAR,1800;
END;
```

Fig. 1   MAP/1 model of an assembly line

modelling constructs. An example of this type of language is MAP/1[5]. In this manufacturing orientated language, the modelling constructs refer to stations, fixtures, parts, and transportation equipment. A MAP/1 model of an assembly line is shown in Fig. 1.

Beyond these simulation languages, there is a support program called TESS[6], which integrates simulation, data management, and graphics capabilities. It provides comprehensive software support for all aspects of a simulation project, including model building, execution, analysis, and presentation. An example of the use of TESS is described later.

To explain better how simulation can serve the needs of the system designer, the next section gives a brief explanation of several simulation studies recently conducted for Carrier Corp. In each case, a description of the system is given, together with the study's objective, modelling approach, and results. The section begins with prefatory remarks on the use of simulation at Carrier Corp.

Applications at Carrier Corp.

As high technology continues its advance into the manufacturing facilities of Carrier, it is increasingly clear that special, sophisticated tools will also be needed to plan, develop, and implement these productivity-improving projects. Carrier has high expectation from these complex work cells and the people who plan and operate them. Issues such as uptime, throughput, reliability, scheduling, etc. are of ever-increasing importance at a time when they are most difficult to understand, quantify, and manage. The 'seat of the pants' approach is no longer adequate.

In the past, simulation at Carrier had been selectively applied only to a few special analyses involving issues such as inventory optimisation and process flow. However, with today's manufacturing orientated languages and simulation support tools, production specialists are now able to perform their own simulation studies on a wide range of planned facility improvements.

Today, simulation is a popular tool for Carrier manufacturing personnel who use it to further improve process performance, especially in the areas of process reliability, productivity, planning, and scheduling. It is currently being used to more quickly and efficiently introduce new processes into the factory environment and to operate them once installed.

Several new process improvement projects were undertaken at Carrier in 1983/84, involving proposed changes in assembly technology and parts processing. Two of these were new systems and one was a refinement of an existing one. Applying simulation to these proposed improvements has allowed Carrier to bracket expected throughput, likely amounts of work-in-process, and needed levels of system component reliability. It has also allowed Carrier to anticipate the impact of these new systems on existing facilities. A discussion of these simulation studies follows.

Case study 1 – Modelling an existing assembly line

Personnel at Carrier's Collierville, Tennessee, plant were interested in examining the impact of implementing several new upgrades in their residential air-conditioning line. But first, they needed an accurate model of the existing line against which to compare these potential improvements. If this could be done, it would add credibility to the results obtained in evaluating these upgrades.

Modelling the existing line[1] consisted of three steps. First, a model was constructed to capture line performance accurately. Secondly, a sensitivity analysis was performed to demonstrate the reaction of the model, and correspondingly the system, to these proposed upgrades. Finally, animations of the line were developed to visually portray the effect of these upgrades.

System description. The assembly line consists of over 75 stations. These stations involve manual assembly, semi-automated testing, repair, and packaging. Generally speaking, the test station areas are the only places where parallel processing is done. A schematic of the line flow is shown in Fig. 2.

Numerous conveyors, both synchronous and non-synchronous, serve as the primary material handling system. All assembly stations are located on power belt conveyors; whereas all test, repair, and packaging stations are located on roller conveyors with controlled stops.

Modelling approach. Inasmuch as the objective of this study was to capture the overall performance of the existing line, only those elements of the system which directly impact performance were included in the model. These included station cycle times, conveyor speeds and distances, buffer capacities, rejection rates at test stations, equipment failure rates, sequencing rules and special processing rules.

A model of the system was built using MAP/1. One of the primary reasons for using this language was its conveyor feature. This feature made it possible, with only a minimum of effort, to capture accurately the characteristics associated with this type of material handling device.

To support the animation requirements, models were also built using SLAM II. TESS, the software package used to develop the animations, uses SLAM II as a modelling medium.

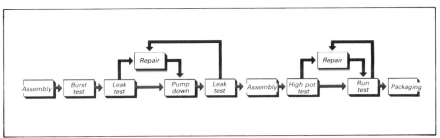

Fig. 2   Collierville's assembly line flow

Results. As previously stated, the first goal was to predict line performance accurately, both overall and by station. As for overall performance, the simulation model predicted throughput within 2.6% of that acutally experienced on the line. As for individual station utilisations, Fig. 3 shows actual versus predicted machine utilisations at a test station. Both the total number of units tested and the distribution of these units across the five machines were predicted with reasonable accuracy. The discrepancies that were noted were due in part to the randomness of certain system occurrences, such as the probability of failure. These, as well as other line performance results, showed conclusively that a simulation model could accurately predict performance, in spite of the system's dynamic nature.

An interesting discovery came out of an investigation into why the fifth machine at this station had such low utilisation. Because it is only chosen if none of the other machines are available, one would naturally expect it to be used less than the others. But why so much less?

By studying the results from the model, it was discovered that this phenomenon was due, in part, to blocking. These five machines at the test station are arranged in a line, with this fifth machine at the end. This makes the fifth machine the first to be blocked if there is sufficient downstream congestion. However, the blocking rarely becomes severe enough to affect the other four machines.

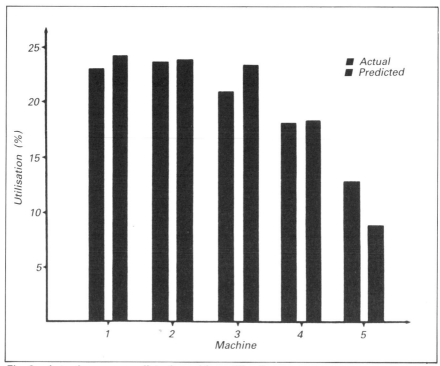

Fig. 3   Actual versus predicted machine utilisation

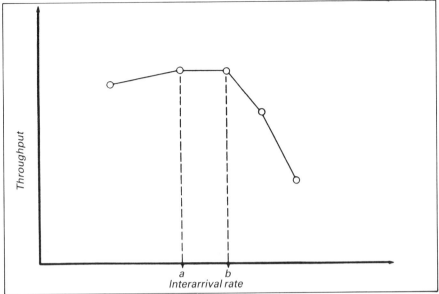

Fig. 4 Throughput versus interarrival rate

One way to distribute better the work load over these five machines is to add buffer following this last machine. In this way, the last machine would not be blocked as often and could, therefore, accept more of its share of the load. Here again, the model could determine how much buffer should be added and what impact adding this buffer would have on overall throughput.

The second step of the analysis was to examine the sensitivity of this assembly system to changes in conveyor speeds. Assuming all assembly operations on the belt conveyors could accept the pace, the system was run at progressively higher speeds.

Fig. 4 shows the system reaching peak production within the interval [a,b]. Furthermore, it shows that decreasing the interarrival rate any more than this would only contribute to system congestion and, accordingly, decrease throughput.

This latter point was clearly conveyed through the use of several animations built in TESS. The five machine test station was one of the areas animated. As the interarrival rate was decreased, congestion in this area built up quickly and eventually restricted the flow of units through this area.

Fig. 5 gives a 'snapshot' of an animation of this test station area. It shows units (small blocks) queueing on the conveyor waiting for one of the test machines to become available. The machines (larger blocks) are colour coded to depict their current state: busy, down, blocked, or idle. As changes occur in the line, corresponding changes occur in the graphic display. For example, when a test machine begins processing a new unit, its colour on the display changes from blue (idle) to green (busy).

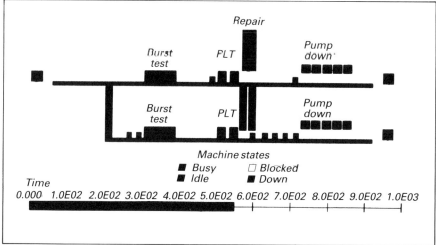

Fig. 5   Animation of test station area (snapshot)

Case study 2 – Modelling a proposed assembly line

BDP, a division of Carrier, is in the process of designing a new assembly line for its Indianapolis plant. To help minimise its capital investment and reduce the number of problems encountered during installation, BDP asked for a simulation study to be done of this line[8]. The primary objective of this study was to determine the minimum station and buffer requirements to produce at the targeted daily production rate.

System description. The proposed line is similar in nature to the assembly line at Collierville in that it consists of both manual and automated operations. These operations include assembly, test, repair, and packaging. There are approximately 65 stations on the line. A schematic of this sytems's line flow is shown in Fig. 6.

Here again, conveyors serve as the primary material handling system. However, this proposed line makes more extensive use of non-synchronous conveyors.

Modelling approach. Although this study dealt with a proposed line as opposed to an existing one, the modelling approach was similar to the one used for Collierville. The main difference was the use of estimated, rather than actual, values. These estimates pertained to such system characteristics as station cycle times, breakdowns, conveyor speeds and distances, and buffer capacities.

It was important in this model to be able to represent station reliability and product quality accurately. With respect to station reliability, the machines at a station were made unavailable for a predetermined amount of time. Typical causes included equipment jams and part outages. As for product quality, failure percentages were used during each stage of a test. If a unit were to fail, it was rejected upon completion of that stage of the test and the station, once emptied,

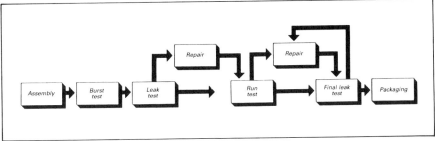

Fig. 6   BDP's assembly line flow

was able to accept the next unit. Thus, the cycle time of a test station was a function of product quality.

One of the main areas of interest with this system was the extent to which automated and manual stations could support each other. Past experience showed that operators at manual stations were significantly affected by the status of their input buffers. For example, as an operator's buffer increased, his assembly time decreased. Thus, the cycle time at manual stations were directly affected by system congestion. To further complicate the design problem, the automated stations could continue processing through breaks as long as there was work available, while the manual stations would stop processing.

To size the buffer requirements properly and balance the complement of manual and automated stations, it was important to capture this mix of work behaviour in the model. The shift schedule and conditional routing features of MAP/1, the language used to model this assembly line, handled these requirements quite adequately.

Results. The model showed that the original design of this line would fall short of the targeted production level by about 14%. Recognising this and the reasons for it, the model was used to test a series of design changes. Options included changing the number of test stations, increasing buffer capacity before and after certain automated stations, increasing the speed of the conveyors, and adding operators. The results of some of these analysis runs are presented in Table 1.

Table 1   Simulation analysis results for five design alternatives

Design alternative	Add test stations Type A	Type B	Add buffer	Increase conveyor speed	Production shortfall (%)
1 (baseline)					14
2	√		√		6
3	√	√	√	√	0
4	√	√	√		0
5	√	√	√ *		0

*Reduced buffer capacity of design alternative 4

Overall, four additional test stations and one extra operator were required to reach the throughput goal. Increasing the speed of the conveyor had no appreciable effect. Additional buffer was also recommended in the vicinity of two of the automated test stations to take better advantage of the 'work ahead' capability of these stations during breaks.

Case study 3 – Modelling a proposed assembly line

Allied Products Division (APD), Knoxville, Tennessee, another division of Carrier, is also designing an automated assembly system. The goal of this system is to produce, at a higher rate, better quality, lower cost units.

A simulation study was done to determine system performance for an assortment of equipment capacities and reliabilities[9,10]. Two design concepts, dial index and assembly cell, were compared to determine which one best met APD's requirements.

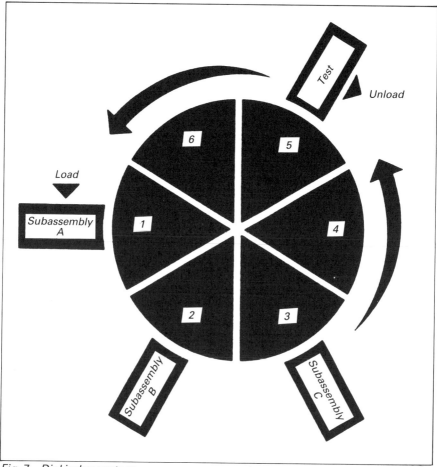

Fig. 7   Dial index system

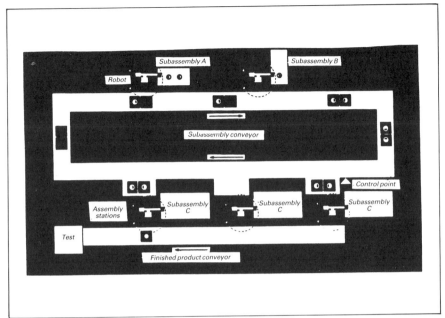

Fig. 8 Assembly cell system

System descriptions. With the dial index concept, a six station table is used. Stations 1, 2 and 3 assemble the unit, Station 4 prepares the assembled unit for removal, Station 5 removes the unit from the table for testing, and Station 6 prepares the assembly fixture for recycling. A diagram of this system is shown in Fig. 7.

The assembly cell concept consists of a closed-loop conveyor which carries one pallet type. This pallet is loaded first with subassembly A, then with subassembly B. Once loaded, the pallet proceeds to a control point where it is sent to one of three assembly cells. The material on the pallet is removed and joined with subassembly C. The pallet is cycled back on the first conveyor, while the completed assembly is sent to the test station via a second conveyor system. A diagram of this system is shown in Fig. 8.

Modelling approach. The general approach to modelling systems in the design stage is to build a model quickly, generalising where possible, without compromising the model's ability to predict performance. This allows for more analysis time and eliminates the often unnecessary and time-consuming task of modelling many of the system's details.

Two simulation models were built for this study, one for each of the design concepts. Both of these models were written in MAP/1.

The system parameters which directly affect throughput are different for these two design concepts. The dial index system is synchronous, meaning all of its stations are directly affected by each other. Here, the cycle times, failure rates, and material supply are the critical parameters. On the other hand, the assembly cell system, being

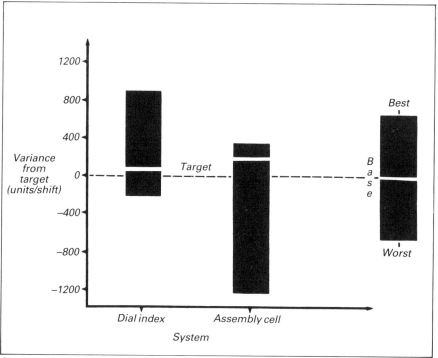

Fig. 9   Performance comparison between dial index system and assembly cell system

non-synchronous, has a different set of critical parameters. While the parameters mentioned above are important to this system as well, it is actually the number of pallets, the conveyor transport time to station cycle time ratio, and the assembly cell buffer sizes that are the most important parameters.

Results. The simulation study revealed that both systems can attain, under certain conditions, the required number of units per shift (see Fig. 9). By varying the data associated with key system parameters (which were different for the two systems), the models were able to reflect each system's probable worst and best performance situations.

In the base case, the dial index system showed a throughput capacity of 69 units per shift above the target. The worst case was 232 units below the target, while the best case was 896 units above the target. The dial index mechanism and Stations 1 and 2 were identified as the critical operations. The study showed that, due to the synchronous nature of the system, changing the cycle times or failure rates of these two stations drastically affected throughput. The study also showed the improvement in productivity one could get by reducing the table size from six to four stations.

The assembly cell system showed a base throughput capacity of 176 units per shift above the target. Because of the distance the pallets have to travel on the conveyor, station utilisation was directly affected by the number of pallets available and the size of the buffers in front of the assembly stations. By varying the number of pallets and the buffer sizes, a worst case of 1252 units below the target and a best case of 336 units above the target were identified. The study also showed that delivering multiple assembly components on a pallet would improve throughput.

Higher base throughput, increased flexibility, ease of expansion, and equipment size are major reasons to favour the assembly cell concept. However, this system carries with it a significantly higher cost. By exercising the simulation model over several design refinements, a determination can be made of the increase in throughput for each incremental increase in cost. Knowledge of the marginal improvements at each refinement stage can then be combined with the other factors mentioned above to determine the best system for APD's needs.

Future of simulation at Carrier

If you were to return to Carrier a few years from now, you could expect to find simulation in expanded use. There will be a continuing effort to use simulation to address critical factory scheduling issues. Capital investment analyses will also benefit from the power of this tool in those instances where high-value, complex systems are being considered. Not only will their performance against plan be an issue, but also how they will fit into the existing system and how they might best be phased into the operation.

System process reliability is also expected to benefit signficantly from a simulation approach. Data gathered in the study will be reflected in process specification to emphasise quantitatively to the vendor just how important system performance is and how it will be analysed and measured by Carrier.

In those instances where a visual model of system dynamics is worthwhile, graphic animation is certain to play an important role. Visual study is expected to reveal issues and relationships not obvious from an analysis of the numerical data. Issues relating to physical layout and control logic will be more vividly described and appropriately studied using this animation technique.

Acknowledgements

The authors wish to acknowledge two individuals for their significant contributions to the work reported in this paper. F. Bradley Armstrong, of Pritsker & Associates Inc., conceived and developed the MAP/1 models used in analysing both the Collierville and Indianapolis assembly lines. He was also the principal analyst on both of these studies. Maged Dessouky, also of Pritsker & Associates, developed the animation of Collierville's test station area.

References

[1] Pritsker, A.B. 1984. *Introduction to Simulation and SLAM II* (second edition). Systems Publishing Corp., West Lafayette, IN, USA.

[?] Kiviat, R.J. et al. 1969. *The Simscript II Programming Language*. Prentice-Hall, Englewood Cliffs, NJ, USA.

[3] Law, M. and Larmey, S. 1984. *An Introduction to Simulation Using SIMSCRIPT II.5*. CACI, Los Angeles.

[4] Schriber, T. 1974. *Simulation Using GPSS*. John Wiley, New York.

[5] Miner, J. and Rolston, J. 1983. *MAP/1 User's Manual*. Pritsker & Associates Inc., West Lafayette, IN, USA.

[6] Standbridge, R., et al. 1984. *TESS User's Manual*. Pritsker & Associates Inc., West Lafayette, IN, USA.

[7] Armstrong, F., et al. 1985. Final Report of the Simulation Analysis of Final Assembly Line #3. Prepared by Pritsker & Associates Inc. for Carrier Corp., Collierville, TN, USA.

[8] Armstrong, F. and Musselman, K.J. 1985. Final Report on the Simulation Analysis of the Proposal Final Assembly Line (LIFE) Project. Prepared by Pritsker & Associates Inc. for BDP Company, Indianapolis, IN, USA.

[9] Martin, L. 1985. Simulation Analysis of the AHED Assembly Cell Concept. Presentation prepared for United Technologies Carrier, Syracuse, NY, USA.

[10] Musselman, J. and Martin, L. 1984. Simulation Analysis of AHED. Presentation prepared for Allied Products Division, United Technologies Carrier, Knoxville, TN, USA.

COMPUTER SIMULATION – A FEASIBILITY AND PLANNING TOOL FOR FMS

R.I. Mills
Ingersoll Engineers Inc., UK

Mainly due to inadequate planning and design, initial flexible manufacturing systems (FMS) did not meet their performance specifications. Simulation, now a proven industrial technique, gives a dynamic insight into proposed systems, and hence it is the ideal feasibility and planning tool for FMS. The technique and its industrial development are described. The philosophy of using multi-level simulation – gaining user experience before system installation, forming the foundations of the system control software, and improvement of the performance of existing facilities – are also mentioned.

To most people the word 'simulation' usually invokes images of an astronaut or airline pilot undergoing training in a specialised and complex machine. These devices are one form of simulation – physical simulation – and are also used for training in other occupations (e.g. train drivers and power station controllers)[1]. This form of simulation may appear very different from that involved in the design of flexible manufacturing systems (FMS), but the principles are very similar, both conceptually and objectively.

The overall objective of training a pilot using a simulator is to minimise both the cost and the risk involved by allowing mistakes to be made in a 'safe' environment, with no risk to life. The objective of manufacturing systems simulation is to minimise the cost and risk involved in an FMS project.

Both the flight simulator and manufacturing systems simulation allow a dynamic computer model to be studied, giving a realistic view of an actual situation. The important point is that the process itself, and not just its output, can be studied.

The obvious difference between these two types of simulation is that a flight simulator can save life, whereas a simulation of a manufacturing system will only save money and time. Hence it is no great surpise that flight simulators have been very highly developed, while the use of simulation in manufacturing industry is still in its infancy.

For a typical FMS project it is essential that all the parameters are fully considered in the planning, so that the final system is as efficient and as operationally sound as possible. Simulation is currently the only effective method available, but it has no capacity to optimise: it can only model a given configuration. Hence an iterative approach is required, with many alternatives being studied. This process can be very time-consuming unless knowledgeable and experienced engineers are available.

There are no real alternatives to simulation which can demonstrate the high degree of confidence required by the high level of investment involved. Conventional planning techniques haved proved unable to cope with the complex and dynamic nature of FMS. In addition, they cannot easily deal with the total systems approach needed to handle FMS successfully. The shortcomings of the early systems in the USA demonstrated how important this systems approach is[2,3].

Technical description

Simulation is an example of the classic scientific method of Hypothesis-Experiment-Analysis-Deduction, which is based on model building. Simulation is one method of evaluating what are called stochastic problems. These are problems where the optimal solution cannot be defined mathematically. Examples of the factors that lead to a problem being stochastic include the non-availability of plant needing to be specified as a distribution rather than as a single average figure, and a need to consider many components and/or a varying mix.

A complete simulation package will consist of the simulation language itself, facilities for display of the results, and optionally a code generator. Obviously, the actual facilities provided vary among the different packages that are available.

Types of language

The three basic types of simulation language are as follows[4]:

Three-phase discrete-event systems. The approach here is sometimes known as the British approach and is used by ECSL, SEE-WHY, SIMON, etc. The three phases are (see Fig. 1a):

- The 'A' phase which advances the clock by the amount necessary to reach the next immediate future event.
- The 'B' phase scans through all the activities currently in progress, terminating those due to finish at the current time.
- The 'C' phase also scans the activities, initiating those which can now be started.

The development of this language type in the UK is shown in Fig. 2.

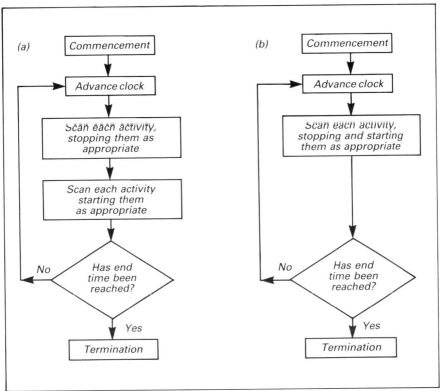

Fig. 1   Simplified flow charts: (a) three-phase systems, and (b) two-phase systems

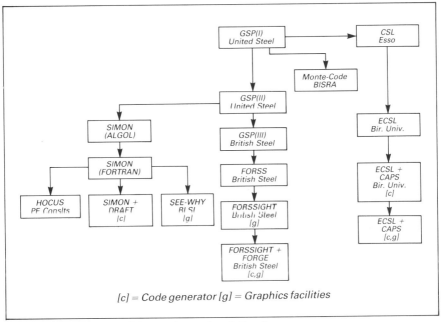

Fig. 2   Development of three-phase simulation languages

Two phase discrete event systems. This approach is very similar to the three-phase system, except that the 'B' and 'C' phases are combined (see Fig. 1b). This means that the ordering of activities is important, whereas this is not critical with a three-phase approach.

The advantage is that the processing is more efficient and quicker, as only one scan through the activities is made at each time-beat. However, it does require significant preliminary analysis of the system.

This approach is usually adopted by languages developed in the USA, such as GPSS, SLAM II.

Continuous systems. These languages use a process-type description of the activities. They are very easy to understand, but suffer from the problems of being difficult to code and to alter. This approach is very little used in industry, except for continuous process plants. The best-known language is SIMULA.

Output facilities

The three main methods of displaying the results of simulation models are outlined below.

Tabulated outputs. The original, and still the most common, method of obtaining results from any mathematical model, including simulations, is in the form of a tabular output of figures. This method is more than adequate for the model builder, but has been justifiably criticised on the grounds that these results are little more than a jumble of figures to anybody not directly connected with the model[5]. Also, the degree of interaction between the model and the user is negligible.

Dedicated physical displays. These were much used during the development of simulation techniques, but have been almost completely superseded by dynamic graphics displays on visual display units. They consist of a mock-up panel resembling the actual control facilities for the system being studied. Varying degrees of interaction can be provided through control dials, switches and gauges as appropriate. Their main current use is for physical simulators for pilot training, etc., as mentioned above.

Dynamic graphics displays. During the last few years the use of dynamic graphics displays, frequently in colour, has become a favoured method of showing the output from a simulation model. This is especially so for microcomputer-based packages, where the refinement of using a postprocessor to control the graphics is often adopted. This shares out the workload and reduces the processing speed penalty imposed by the use of graphics. Often some degree of interaction is provided allowing some parameters (e.g. batch sizes) to be changed by a user (as distinct from the model builder).

This method of presenting a simulation is beneficial, provided that it is remembered that the basic model is not altered: it is only communication of that model that is improved. Accordingly this technique is at its most useful for 'selling' a concept, for gaining the confidence of local management, and for training purposes.

Graphics do have some serious disadvantages that are frequently not considered. They can be time-consuming to write and run, and can cause the concentration of attention upon the effect(s) of a problem, rather than on the problem itself. Overall, the availability of these sophisticated graphics packages is a major advantage for easy communication, but they place an additional responsibility on simulation modellers to ensure that these facilities are not misused.

Code generators

These are computer programs which enable the input of the problem in an interactive and English-like question and answer format. Computer simulation models are then written by these programs. Some simulation packages include a code generator; others have one available as an extra.

Code generators vary in the level of complexity that they can handle, but for a very complex or intricate model it will be necessary to 'patch-in' hand-written pieces of code.

The advantages are that they enable the writing of simple models with no direct use of the computer language, and that they reduce the tedium of code-writing with more complex models. Hence their use can speed up the construction of almost all models. Their disadvantage is that they tend to be pedantic in operation, and it is difficult to bypass any section. This can make the writing of repetitive models very time-consuming, and in these cases it may be easier to use the actual language.

History of industrial development

Although the principles of simulation have been known for some time, it was computer power that made it a viable technique. The first genuine computer simulation package was written at United Steel by Dr K.D. Tocher and his team. This package, called the General Simulation Program (GSP), was announced in 1958[6]. It was an advanced package and has provided the basis for almost all the three-phase discrete-event languages in use today. Development continued in the iron and steel indsutry with GSP leading to GSP(II), GSP(III) and currently FORSS. Work in the petrochemical industry resulted in the development of CSL at Esso. These two industries were also the major users of simulation, typical uses being to establish the maximum refinery storage capacity required, and to compare the operating rules used in steel mills.

Improvements in computer techniques allowed computer software, including simulation packages, to become more 'portable', i.e. able to be used on more than one make of computer. This feature increased the usefulness of such packages, leading to their greater use, especially in universities, where some manufacturing system problems were studied.

This feature also permitted the transfer of simulation packages to mini- and microcomputers as these became available.

The introduction of flexible manufacturing concepts and the resulting increased complexity of manufacturing systems has led to simulation becoming much more widely known and accepted. This is especially true in the USA[7] and West Germany[8]; currently little is known about its uses in Japanese industry. More recently the technique has become accepted in the UK with industrially linked projects in British universities[9], and its use during the design phase of FMS projects, such as SCAMP[10].

The industrial use of simulation is bound to increase, driven by the need to understand even more complex manufacturing systems. Leading software houses have recognised this potential and are marketing their simulation packages with regard to FMS simulation.

A philosophy for the use of simulation

The objective of this philosophy is to provide a flexible framework for the use of simulation in a real project. The method has been proved out during feasibility studies. The aim is to provide the maximum amount of information during the early stages of a project, without the need for the commitment of large resources. The key element in this approach is the medium-complexity modular models, which fill the gap between general-purpose 'systems simulators'[11] and full-scale detailed simulation models. This level of complexity is crucial to the industrial user and yet so far has been paid very little attention.

Use of language types

For all the work with a specialised simulation language that is described in this paper, the Extended Control and Simulation Language (ECSL) was used. This three-phase discrete-event language was chosen because of its availability, its portability, and its useful code generator, which is known as CAPS.

The modular models written in a high-level language use a variant of the two-phase system to improve processing efficiency, and so the ordering of the activities in these models is important. This approach was used simply to minimise the processing times, a factor of special importance as these models were developed on a microcomputer with accordingly restricted computing power.

Construction of a simulation model

The construction of a typical model follows four principal stages:

- Define the problem – A basic examination of the problem will frequently highlight the critical areas and attention can then be concentrated on them.
- Establish and collect the data required – This is the most difficult and time-consuming part of the process. The amount and detail of the

data required must be decided and its collection arranged. This can be extremely difficult for a proposed system, and frequently approximations or estimates have to be used. Obtaining realistic reliability statistics for modern plant is also difficult[12].

- Build the model – This is comparatively easy, given the appropriate training and some experience of the language.
- Model validation – The process of ensuring that the model does what is wanted and in the way that is wanted. This is usually done by comparing several runs of the model with each other and with results obtained by other methods. Again, this is more difficult for proposed systems where comparisons cannot be made with the 'real world'. The model should be checked again after any alterations have been made.

Following the completion of the simulation model, it can be used. Several runs should be made at each stage, then changes can be made to improve its performance, and so on, in an iterative manner. The use of several runs at each stage helps to eliminate the random effects that can occur on small samples from statistical distributions. Short-cutting this procedure can cause serious errors to occur.

Levels of simulation

Several different types of simulation model will be found useful at various stages in an FMS project. The relationship between them is shown in Fig. 3[13]. These different types of simulation vary in their complexity, in the amount of data required, in the type and size of computer needed to run them, and in the scope and accuracy of the results that are obtained. A compromise between these has to be established, and this will depend on what is required at any stage of a project. A summary of the main model types and their salient features is given as Table 1.

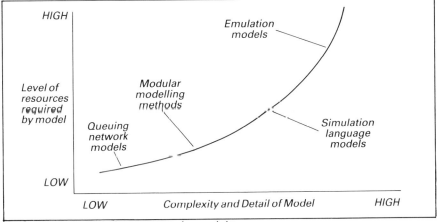

Fig. 3   Relationship between main model types

Table 1   Summary of main model types

Model type	Requirements			Outputs	
	User skill	Computer type	Quantity of data	Level of detail	Level of confidence
Queuing network	low	micro	small	low	fair
Modular methods	fair	small mini	fair	fair	high
Simple simulation language	high	small mini	fair	fair	high
Emulation	high	mainframe/ super-mini	very high	very high	very high

Queuing network theory based models. This type of model is not strictly simulation, being based on standard queuing network theory. They have the major advantage of fast response, allowing rapid evaluation of many different options in the early stages of a project. Their main disadvantages are the limited level of output provided and its relatively low level of accuracy. In addition, their theoretical foundation is complex, and this tends to reduce user confidence due to a lack of comprehension as to how the model operates.

One special disadvantage is that blocking – the effect of components competing for an insufficient quantity of resources – cannot be correctly modelled and a mathematical 'fiddle' has to be used. As blocking is probably the most serious factor affecting FMS performance, this is potentially a serious limitation.

Models can be written directly, but this is laborious and time-consuming, as standard software packages are available, such as CAN/Q[14] and GMS[15]. These packages can be run on a microcomputer and are easy to operate.

Because of their ease of operation, these models tend to be overused. They are designed to indicate general trends and no confidence can be placed in their ability to reflect accurately the effect of small changes in their input data.

High-level-language modular models. These models consist of modules or blocks of high-level language code (i.e. FORTRAN, BASIC), which can be rapidly built up to represent the desired hardware configuration. A typical module will represent a machining station or a section of a conveyor.

Although these models require a greater degree of input resources in terms of model construction and processing times, they have the advantage of providing more output, which is of a higher degree of accuracy. In addition, they can be tailored to suit a given requirement, with breakdown statistics, special hardware configurations, etc., able to be incorporated into the model. User confidence in the model is high as the principles of its logical construction are easy to grasp.

The disadvantages are that the models are still simplifications and all the complexities of a real system cannot be handled, and that better computing facilities are required. Although this type of model can run on a small microcomputer, this incurs a penalty of increased processing times, and ideally a minicomputer should be used to give acceptably short turn-round times.

Simple simulation language models. These are models written in a language specially designed for simulation. They have the advantage of very efficient use of computer time, due to the specialist simulation facilities provided in the language. Additionally, they can form the basis for the development of more detailed models.

This type of model has to be written from scratch, which is a lengthy process requiring highly skilled personnel. Although the recent availability of code generators has reduced the amount of typing required, a thorough understanding of the system being modelled is still required.

It is unlikely that a model of this sort would be attempted unless it was envisaged that an emulation-type model would be required at a later stage.

Emulation models. These are the ultimate development of the simulation concept as currently applied to manufacturing systems. They are inevitably written in a simulation language and usually developed from simpler models. They require an enormous input – they can take up to two man-years to write, contain upwards of 2000 statements, and impose very heavy demands upon computing resources. A mainframe computer, usually operating in a batch mode, is the only practical computer configuration for this type of model.

In return for this investment, these models provide a detailed insight into the complete system, to such an extent that they can be used as the foundations for the control software of the finished installation. This type of model can only be justified when the layout and overall framework of a proposed system has been decided. They can then be used for fine-tuning of that system and for comparison of alternative strategies for dealing with breakdowns, product mix changes, etc. These models should be maintained so that changes to a running system can be tried out before they are implemented.

The philosophy in practice – An industrial example

An industrial example will be used to demonstrate the uses of the different levels of simulation. Particular attention will be paid to the modular models, as these form the cornerstone of the philosophy. The system to be described was modelled as part of the feasibility study for an application to machine three families of castings, required in approximately equal quantities. At this stage no attempt was made to

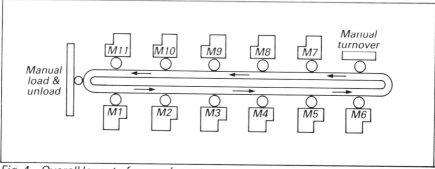

Fig. 4   Overall layout of example system

the basis of three different components only. The system layout is given schematically as Fig. 4.

Use of queuing network models

Initially the model used was CAN/Q[14]. Although simple, this model showed that the system should be adequate for the output required, but the utilisation of the machine tools was unbalanced. The model also showed that the conveyor system could easily cope with the demands made upon it.

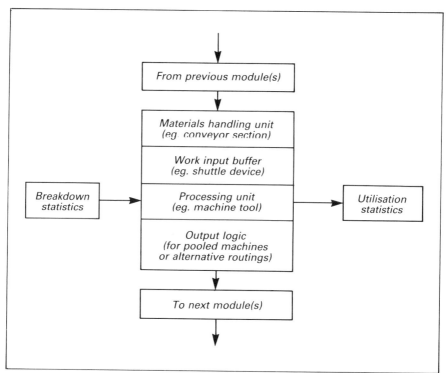

Fig. 5   Basic arrangement of one block of example modular model

Development of a high-level-language modular model

Earlier work within BL Technology (now Gaydon Technology Ltd) had established the potential of microcomputers (such as a CBM 4032) for fast-response simulation work[16]. Building on this work and with regard to the need to use this approach for other feasibility studies, a modular model was written in BASIC. This was initially developed on a CBM 4032 computer, but was later transferred to a VAX 11/780 mainframe to give greater processing speed. Although the model was initially developed for the one system, it has since been generalised. This method thus has potential for more general use.

The model consists of a series of blocks (each representing one processing unit and part of the materials handling system) between which components are moved as required. The basic structure of a module is shown as Fig. 5. In practice the writing of a set of modules from a standard format is a fairly fast process. Once these modules have been written, the model is completed by the addition of standard blocks covering input and output, a controlling executive, and specially written statements to cover the system scheduling and priorities.

Typically, this complete process should take 2-3 days for the modelling of an average-sized FMS.

Use of the modular model

Originally, each station had a four-station rotary table, and the operations were assigned to pairs of machines as shown in Table 2. Subsequent stages pooled machines in larger groupings. An improvement in performance was expected, but only slight differences were

Table 2   Summary of the results of models

Model type	Machine pooling											Shuttle type	Break-downs applied	Outputs	
	M 1	M 2	M 3	M 4	M 5	M 6	M 7	M 8	M 9	M 10	M 11			Pro-uction rate (per hour)	Through-put time (min)
Queuing network												–	–	95	63
												–	–	98	61
Modular high-level language												4-station	none	113	45
												4-station	none	112	45
												4-station	none	113	45
												2-station	none	114	44
												2-station	O/P × 85%	97	–
												2-station	randomly	92	49
Simple simulation language												2-station	none	122	49
												2-station	randomly	101	51

noted. In addition, the pooling of machines 1–4 caused an increase in
the size of the tool magazine required, for a very marginal increase in
output. The replacement of the four-station rotary tables by less costly
two-position shuttles was tried, again with only a slight effect on
performance.

The effect of breakdowns was investigated by comparing the
traditional method of multiplying the full-rate output by 85% with a
simulation based on randomly applied downtimes. These two methods
gave different results. In this case the differences were relatively small,
but it started to show the problems of using conventional techniques
with complex systems. The relatively good showing of the queuing
model is notable.

The effects of the alternatives mentioned on the system performance
are summarised in Table 2. Many other options could have been
considered.

Models written in a simulation language

The next step in the planning process is the writing of a simulation
language model. This can then be developed as required according to
the progress of the project. For this feasibility study, only the initial
model was written. Its results (also shown in Table 2) were found to
correlate very well with those of the latest version of the modular
model. The better modelling of the system control logic allowed by the
more complex algorithms used accounts for the increased output.

Use of the results obtained

Confidence building

Simulation, as well as providing the optimum method of comparing
alternative system approaches during the design stage, also enables
increased confidence to be developed, at all levels, in the success of the
project.

A dynamic graphical simulation model, especially in colour, is more
readily acceptable than sets of figures – it seems to be a real system,
rather than an abstract concept. Hence the use of such a model gives the
following benefits:

- 'Selling' the concept to management – the objectives are to show the
  thoroughness of the planning undertaken and to demonstrate how
  the finished system will work.
- The use of the model by the project team to investigate any potential
  bottleneck areas, as well as to develop different control and
  scheduling strategies.
- The personnel who will operate the system can be given 'hands-on'
  experience using the system model. This can reduce the time spent in
  commissioning the system.

System control foundation

A completed emulation model will include the executive (or controlling) level logic required to run the system. This model can thus be used as a 'proving ground' for this logic, which can then be written into the system control computer. This procedure will reduce the number of control software faults to be dealt with during commissioning.

This may appear to be the obvious route, but system simulation and the writing of the control software are often undertaken by different sub-contractors, and so this useful short-cut is not fully used.

Other uses

As well as assisting in the development of new facilities, simulation can also be used to help improve the performance of existing facilities. This is a less demanding application as most of the data needed can be obtained from the current situation. The method used is to build a model of the current system, which can then be thoroughly validated using the current situation. This model is then used to prove the soundness of the approach, before the proposed improvements are included. The effects of these changes can then be assessed and decisions made on their implementation.

This approach has been used on some of the early FMS installations in the USA to increase their performance, especially their output. However, almost any type of improvement to a manufacturing system can be 'dry-run' using simulation.

Obviously, this approach is only justifiable when a complex system is involved, when it provides a convenient, inexpensive and fast method of planning changes.

The future

Current research trends

There are three main areas of research within the field of manufacturing system simulation. All of these are concerned with FMS, as this currently offers the greatest potential for the use of simulation:

- The use of simulation to produce general guidelines for the design and control of FMS. This work aims to generate a set of basic rules for the layout of systems and their control logic. This should enable the design process to be streamlined as the first layout and control system will be nearer to the ideal. Also, it is likely that these methods will considerably shorten the iterative process that has to be followed to obtain the best practical arrangement for any given set of criteria.
- The development of low-cost fast-response simulation methods that are easy to use and understand. The modular modelling technique is one of these methods.
- The development of the present simulation languages to include

better user facilities, especially for the less experienced user. Two of these topics are code generators, for easier model writing, and colour graphics, intended to improve communication. This work is mainly being carried out by software houses seeking to improve the acceptability of their microcomputer based simulation packages.

Future possibilities

There are many possible developments in this field. Some of the more likely ones are:

- The incorporation of high-resolution dynamic graphics, similar to those used by a CAD system. This would enable the construction of a moving picture including robot movements, etc. This would be an aid to 'selling' a system concept, as well as easing system analysis for non-computer-biased personnel.
- Simulation software will become more 'user-friendly' allowing easier use by non-specialist personnel. Greater computer power will enable reduced processing times.
- A change in the method of structuring simulation models so that the control and scheduling statements are included in an executive level of the model. This section could then be lifted out and used as the top level of the system control software. Hence new control software could be developed and proved out without disrupting production. Direct down-loading of this software to the system control computer would maintain a high degree of confidence in its functional integrity.

Concluding remarks

The complex nature of an FMS, coupled with the need to maximise its performance, means that such a system can be effectively designed only by iterative computer simulation methods. This provides the only way of understanding these systems in their entireity, thus avoiding costly mistakes.

The philosophy of using different models of increasing complexity during the various stages of project planning enables the simulation tool to provide the necessary information at the right time.

Simulation, particularly when using colour graphics, aids communication and builds confidence in a project, by allowing potential problem situations to be recognised and the appropriate corrective action taken. This allows management to quantify the risks involved, the project team to understand the system constraints, and the operating personnel to see its functional ability.

It will have become obvious from this paper that personnel competent not only in machine tool aspects but also in engineering computing are required. The lack of such people will be a greater restraint on the use of FMS than the adequacy of the simulation methods available.

Overall, simulation is an essential tool for the feasibility and planning of all complex manufacturing systems, but especially FMS, which is still a step into the unknown for most companies.

References

[1] Philpotts, P. 1982. Flights of fancy. *Computer Systems*, March 1982: 41-45.

[2] Ingersoll Engineers, 1982. *The FMS Report*. IFS (Publications) Ltd, Bedford, UK.

[3] Bryce, A.L.G. and Roberts, P.A. 1982. *Flexible machining systems in the USA*. In, *Proc. 1st Int. Conf. on Flexible Manufacturing Systems*, pp. 49-70. IFS (Publications) Ltd, Bedford, UK.

[4] Martin, F.F. 1985. *Computer Modelling and Simulation*. John Wiley & Sons, New York.

[5] Hurrion, R.D. 1978. An investigation of visual interactive simulation using the job-shop scheduling problem. *Op. Res. Q.*, 29(11).

[6] Tocher, K.D. et al. 1985. Simulation studies of industrial operations. *J.R. Stat. Soc.*, A122(4): 484-510.

[7] Barton, J.E. 1980. Try simulation for sizing up systems – Before its too late. *Production Engineering*, December 1980: 42-45.

[8] Warnecke, H.J. and Vetting, G. 1982. Technical investment planning of flexible manufacturing systems – The application of practice-oriented methods. *J. Manufacturing Syst.*, 1(1): 89-98.

[9] Rajagopal, K. 1980. Ph.D. Thesis. Department of Mechanical Engineering, UMIST, Manchester, UK.

[10] Rathmill, K. and Chan, W.W. 1983. What simulation can do for FMS design and planning. *The FMS Magazine*, 1(3).

[11] Bevans, J.P. 1982. First choose an FMS simulator. *American Machinist*, May 1982.

[12] Stewart, E. 1976. A survey of breakdowns of NC machine tools, MTIRA Research Report No. 67.

[13] Rathmill, K. 1979. Interactive computer modelling provides a new laboratory facility for production systems. *I.J.M.E.E.*, 7(3): 147-150.

[14] Solberg, J.J. et al. 1976. The optimised planning of computerised manufacturing systems, Report No. 3, School of Industrial Engineering, Purdue University, IN, USA.

[15] Kay, J.M. and Walmsley, A.J. 1982. Computer aids for the optimal design of operationally effective FMS. In, *Proc. 1st Int. Conf. on Flexible Manufacturing Systems*, pp. 463-480. IFS (Publications) Ltd, Bedford, UK.

[16] Hallstron, I. 1981. M.Sc. Thesis. Department of Production Engineering, University of Aston in Birmingham, UK.

2

Risk Evaluation –
Improving the Odds of A Successful
Manufacturing Enterprise

The simulation technique may be used through the complete life-cycle of designing, installing, controlling and operating a manufacturing system. This section considers the role of simulation as a method of helping to quantify financial decisions and their associated risks at the different stages of a manufacturing system project. It also highlights the integrating nature of a simulation model so that it is possible to concentrate on individual components of a manufacturing system and still see their effect on the whole system.

SIMULATION AND REDUCTION OF RISK IN FINANCIAL DECISION-MAKING

J.F. Wilson
Ingersoll Engineers Inc., UK

The practice of applying manufacturing technology in a piecemeal way is being replaced by the installation of large, complex flexible manufacturing systems or computer-integrated manufacturing systems. These systems represent a major investment for any size of company. Flexibility is the key to receiving long-lasting benefit from such investment. Flexible manufacturing allows companies to counteract the effects of unpredictable external variables such as changes in the market, availability of skilled staff, supplies of raw materials, financial restrictions and competitors' activities. The risks involved in making financial decisions are defined and the use of simulation models both to quantify and to reduce these risks is described. The discussion also identifies the relationships between the cost and timing of simulation modelling, and relates these to the scale of costs at each stage of a major project, from inception to commissioning.

The practice of applying manufacturing technology in a piecemeal way is being replaced by the installation of larger and more complex systems. Market pressures create a sense of urgency about making important investment decisions.

Simulation modelling can offer the key to better systems because it improves the quality of information on which to base decisions, but it can be expensive and time-consuming. So when should simulation be used and what form should it take? To determine that the following questions need to be considered:

- How vulnerable is the business to external forces?
- What is the level of risk involved in manufacturing investment decisions?
- How can simulation reduce these risks? Can simulation techniques be trusted?
- Is simulation good value for money?

Simulation in perspective

Before discussing the above factors in more detail, it will be useful to compare simulation with other more familiar aids to systems planning.

In the planning of a complex installation, such as a flexible manufacturing system, a two-dimensional layout is normally produced. This shows the position of machines, processes, storage, and type of transportation. It helps explain and visualise the proposed system, but does not demonstrate how it will work. A three-dimensional model does the same thing, but with much greater clarity.

If a 3D model is made to work, then its behaviour can be studied to identify bottlenecks and anomalies. The model can be refined and extended until optimum performance is achieved. Such models have been tried, operated either by moving its elements manually like a war game or automatically by using small electronic motors under computer control. However, working models are expensive to build, conducting experimental programmes can be difficult, and it takes a long time to obtain results.

With the use of a computer, it is possible to create an electronic model of the proposed manufacturing system and to study its behaviour as moving colour graphic displays on a viewing screen, and using print-outs of statistics generated during experimental set-ups. When the model is operational, the scope for experimentation is vast, and results are obtained very quickly.

Simulation is not a technology that stands on its own: it is an integral part of project planning. It comprises:

- Definition of the proposed system installation, including control and scheduling functions.
- Writing and proving of the electronic model.
- An experimental program accompanied by development of the installation concept, particularly the control and scheduling algorithms.

External pressures on manufacturing industry

If UK companies are to survive, they must face up to international competition. It is no longer good enough to make small incremental changes to improve efficiency. Only aiming to be the best will ensure success, and so major investments have to be made. Long-lasting benefits from these major investments will only be obtained if the new systems are flexible enough to:

- Adapt to unexpected changes in market demand.
- Function in spite of fluctuations in availability of materials.
- Be operated by people with various levels of skills.

Flexible manufacturing systems (FMS) and computer-integrated manufacturing predominate in investment strategies because they provide the flexibility necessary to compete in international markets

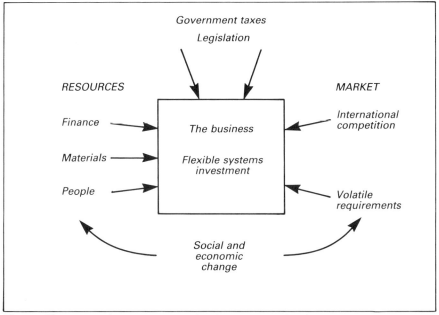

Fig. 1   The external pressures on a business promoting large systems investments

(Fig. 1). These new manufacturing systems generate vital business opportunities, and without them many companies will fail. Timing and accuracy in decision-making are therefore critical and risks are high.

The high risks in systems installations are:

- Large system installations will radically change and disrupt the company's operations.
- The combination of machines, storage systems, transportation, and the use of computer technologies is unfamiliar to many people.
- System performance depends not only on the performance of the individual elements, which can readily be predicted, but also on the interaction effects between them, which cannot be analysed by conventional manual methods.

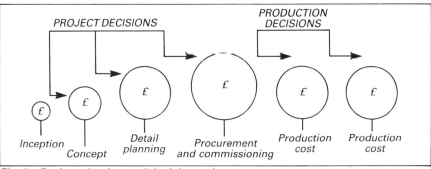

Fig. 2   Project phasing and decision points

In practice, major investments are made as the result of a series of decisions. At any stage, a negative but incorrect decision will deprive a company of an opportunity that may be critical to its survival, while a positive but incorrect decision will waste resources. What is needed is clear and precise information so that decisions can be both accurate and timely. The main decision points in an investment project are shown in Fig. 2.

As a project progresses from inception to production, each decision point commits a further increment of expense. These escalate, with the largest being procurement. Subsequent decisions on production strategy can also be of great financial significance, as they will affect both production output and profitability. Each project step is in preparation for the next and so the expense involved should be reasonable in relation to the cost and risks of the next stage.

Simulation models can be used to obtain advance information on system performance, to refine the layout and size of the elements within the system, and to develop control logic during any stage of a project. But the level of detail in the model, and therefore the expense involved, should be reasonable in relation to the degree of risk and level of expense resting on the next decision.

Project inception

Ideally, projects should be controlled by a top-down approach, based primarily on market requirements, followed by a business plan and then a manufacturing strategy. Much of the data for the justification to proceed with a project concept study will then already have been prepared as part of this company-wide study.

Project proposals generated in a more random way by individual champions of new technology, such as suppliers, will require considerably more effort in collection and collation of information to make the case to proceed.

A concept study for a project of £1 million may cost £25,000–£50,000 at a true costing of resources used and so a decision to proceed cannot be made lightly. It is not usual at this stage to support the decision with detailed analysis or simulation because little basic data will be available. Larger projects, however, may justify a pre-feasibility investigation. This could usefully be supported by simple low-cost simulations based on queuing theory. These models handle average rather than random distribution values of input parameters and so generate approximate information for sizing a proposed system.

Concept study

The purpose of the concept study is to determine the most suitable system, its cost, the time required to enter production, the systems performance and the risks involved. The main purpose is to quantify these factors so that a decision whether or not to implement can be

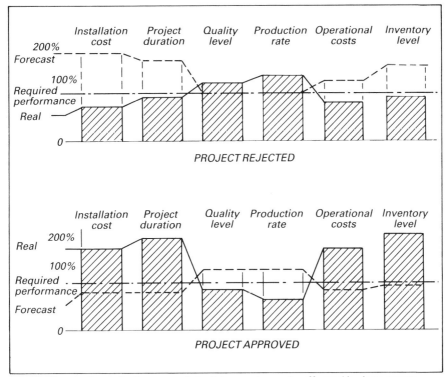

Fig. 3   Example project approval decisions adversely affected by inaccurate information

made. The wrong decision could have a major impact on the company's competitive position. Any of a number of inputs of information may be wrong, as illustrated in Fig. 3. A valuable project may be rejected or a 'white elephant' may be accepted because of faulty or inadequate information, although the latter may be spotted at the detail project planning stage.

Lack of information can also delay approval, and this can have two adverse effects. First, the project may not be completed in time to take advantage of the market opportunity; and secondly, production commitments may have to be met using inefficient methods or by expensive bought-out capacity. This is illustrated in Fig. 4.

Dynamic simulation models are required during the concept study to determine the best layout, to size system elements accurately, and to establish the flexibility and output of the system in response to different order entry patterns. Its purpose is to determine space requirements and investment costs and to confirm that performance requirements can be met.

It is not necessary at this stage to model the method of transportation within the system, its work scheduling or control logic. It is usually acceptable to rationalise some of the input data to simplify the model, thus reducing costs and saving time.

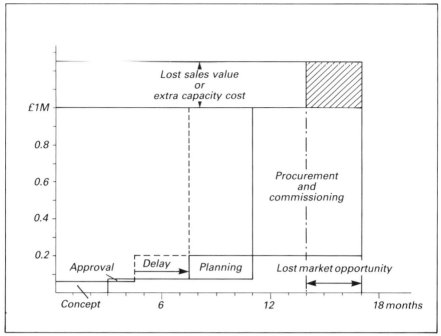

Fig. 4   Example of effect of approval delay on project

Project planning

In this phase of a project every detail of the system must be engineered and precisely specified in preparation for procurement and commissioning. The purposes are to ensure that the system will perform as required and to adjust the estimate of project investment cost, if necessary.

Lack of analytical methods to determine the interaction between the elements of a complicated system makes the use of simulation modelling essential. Without it, the performance of the system cannot be predicted with any credibility.

Using simulation during project planning will help to avoid a number of risks, such as:

- Specification of unnecessary equipment, thereby wasting money and space.
- Failure to specify some essential items, leading to extra expense and delay during commissioning, when these faults will have to be specified.
- Need to locate and pay for temporary manufacturing capacity in order to supply customers, or to accept a loss in sales.

These are illustrated in Fig. 5.

The greatest risk of all, of course, is that the system performs so badly that the investment is wasted and the business opportunity is lost.

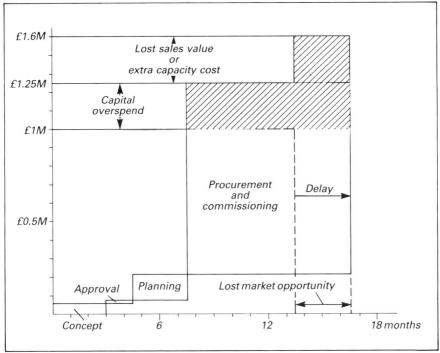

Fig. 5   Example of effect of inaccurate planning on project

At the project planning stage, the dynamic simulation model will include transportation, scheduling and the control logic of the system. These aspects are needed to explore the operation of the system fully so that it can be specified accurately. The scheduling and control logic handling within the model are particularly significant in reducing software costs and reducing the risk of delays in project commissioning. Advance trials of the logic within the model permit more accurate specification of the system's software. The supplier's task is simplified and the behaviour of the computing system becomes more predictable.

For very large projects, implementation may be carried out in successive phases. The reasons for this approach include:

● Minimising the disruptive effects on the organisation.
● Matching system capability with market demand.
● Balancing the availability of investment monies.

Each phase is in effect a new system and its performance can best be predicted using simulation models which, with forward planning, can be expanded as each phase is added.

Procurement and commissioning

The relatively long time required to procure, assemble and commission all the various parts of a system provides the opportunity to prepare for production operation. Even though the installation may function perfectly under test conditions, it still has to be started up, manned, supervised, maintained and so on. It will also break down at times and therefore contingency plans are needed. The system will have to handle new and varied work, possibly requiring modifications to the system. If people are not suitably trained and procedures not planned to cope with all these operational aspects of the system, the required production performance will not be achieved.

The simulation model used to assist project planning can be used, possibly with further refinement, to prepare for production. Experimental programs can be used to resolve the following:

- Procedure and time required to start up production and to close down to a standard condition ready for restart.
- Behaviour of the system and procedures to be used when parts of the system break down or are withdrawn for maintenance.

Supervisors and managers can use the model to learn about the system before it comes into operation. Subsequently, simulation can be a permanent feature of production. Continually updated with performance statistics monitored during operation, the model can be used to verify advanced loading schedules and evaluate the future effects of proposed production decisions. These may include, for example, modifications of the system and changes in product requirements.

Concluding remarks

Simulation modelling techniques will substantially reduce the risks in decision-making throughout the course of a project to design and install a large manufacturing system. The level of modelling detail and accuracy, and therefore the cost and duration of the work, should be commensurate with the investment level, business implications and risks involved in each decision to proceed to the next project phase.

References

[1] Rathmill, K. 1979. Computer simulation of the factory of the future. In, *Proc. 63rd Infotech State of the Art Conf.*, Amsterdam, October 1979.
[2] Browne, J. and Davies, B.J. 1984. The design and validation of a digital simulation model for job shop control decision-making. *Int. J. Prod. Res.*, 22(2).
[3] De La More, R.F. 1982. *Manufacturing Systems Economics*. Holt, Rinehart and Winston, London, Chap. 5, pp. 71-93.
[4] Solberg, J. 1979. Computer models for design and control of flexible manufacturing systems. In, *Proc. 16th Annual Meeting and Conf. Numerical Control Soc.*, Los Angeles, CA, March 1979.
[5] Dupont-Gatelmand, C. 1983. Key success factors for FMS designs and implementation. In, *Proc. 2nd Int. Conf. on Flexible Manufacturing Systems*, pp. 283-294. IFS (Publications) Ltd, Bedford, UK.

RISK AVOIDANCE THROUGH INDEPENDENT SIMULATION

P.L.C. Dunn
GEC Mechanical Handling Ltd, UK

Flexible manufacturing systems (FMS) represent an evolving technology, the concept of which is now widely understood. In parallel with the scientific development of FMS technology (although more usually ahead of it) are the claims of system suppliers. It is rapidly becoming evident that an independent simulation capability is a fundamental requirement when venturing into the marketplace for an FMS. The associated risks are examined and different methods are suggested by which a potential FMS user can select a suitable simulation language or program. Some advantages and pitfalls created by this approach are discussed.

The implementation of advanced technology in the form of an FMS can, in many instances, represent a significant risk to corporate profitability – or even survival. The principal areas of risk are readily identifiable:

- Total project cost.
- Project timing.
- Functionality.
- Changes in product volumes or mix.
- Product range expansion or contraction.

The primary objective of this paper is to identify the risks in isolation and in combination and to discuss how to select a suitable simulation tool to reduce or eliminate these risks. Both in-house and external sources of simulation expertise will be compared in general terms since it is important to realise that no universally acceptable solution exists. That does not mean that a single simulator cannot be used to represent a very wide range of different situations, since many do. However, the methods utilised for model creation and the type of hardware employed are today the main criteria for selection of a general simulation strategy. The current trend is towards encouraging production engineers, not programmers, to design manufacturing systems.

Corporate risks

Total project cost

This is very difficult to determine and will always encompass a degree of estimation depending on:

- Product types.
- Manufacturing methods.
- Types of machinery.
- Material handling requirements.
- Degree of integration.
- Sophistication of software.
- Inspection, tooling control, etc.

As a general rule, the cost of such a project will start at the level approved in the capital budget and increase from there due to oversights or a variety of justifiable additional requirements.

Project timing

Overall system implementation timing is based on estimates and assumptions that a certain level of resources will achieve the required results in a given time. Again, the time required to achieve a goal will normally expand to fill or overrun the time available. Late deliveries of machine tools, software or indeterminate critical items will only serve to extend this time.

Functionality

The previous two elements pale into insignificance when functionality is considered. This is the overriding factor governing both total cost and timing since until functionality is achieved the grand total cost and project timing cannot be known in real terms.

Product volume/mix changes

Such changes could either spell disaster or be easily accommodated depending upon the foresight applied at the conception/design phase, the potential of the system for expansion or contraction, and perhaps the availability of alternative products or machine tool hardware.

Product range changes

Here again, the impact of such changes is dependent upon the factors discussed in the previous paragraph.

General risk avoidance

Project planning

Unless a high level of accuracy can be guaranteed in terms of identifiable products, product life-cycles and volumes, then flexibility must be the key word. It is of course possible to define flexibility, but it will soon become apparent that the ultimate in flexibility cannot be justified. For example, if current requirements indicate a need for milling, drilling and turning, it would be unrealistic to include gear cutting, grinding, sawing and honing machines just to be flexible. The creation and maintenance of a formalised project plan is mandatory as a basis for project control.

System functional specification

Whichever route is taken to arrive at a fully operational system, it is essential during the project planning stage to formalise the system specification. Without a detailed statement defining the component envelope, production, process parameters, facility requirements and acceptance criteria, it will not be possible to measure the performance of the system or to gauge whether the system funded is the system obtained.

The 'safe approach'

This label is used here to describe the widely adopted philosophy of specifying a high-level FMS and then reducing the technology until comfortable with the concept. This approach seldom aims to reduce the level of hardware within a system – invariable hardware is increased to gain 'overkill confidence'. "We can always resort to manual override when (not *if*) something goes wrong." This may appear to be a valid method of reducing the risk element in a project and perhaps also of reducing costs. It may not, however, be pratically sound or financially viable due to the origin of the primary returns which are generated by the control software of the system.

Computer simulation

Computer simulation should be considered, not to verify the operation of a system already designed, but to provide a basis and a proving ground during the preliminary design process.

Computer simulation

This paper provides an insight into the role of simulation, the concept of a computer model an the application of simulation as a design tool to eliminate or reduce the effect of the risks identified.

The computer model

The computer model may be defined as a software representation of the elements of a system, and of structural or physical relationships and time-based interactions within that system.

The role of simulation

To be effective it is necessary to employ simulation techniques during the design phase of an FMS and to be able to evaluate alternatives. The design evaluation process will examine hardware items:

- Parts.
- Machines/stations.
- Pre-, in-, and post-process storage.
- Material handling/routing.

and software control conforming to assumptions and program rules:

- Part balance/schedule.
- Operation sequences.
- Station sequences.
- Storage and queuing control.
- Transporter selection.
- Transporter control.

Execution of the simulation program will result in the generation of measurement data against which the following performance evaluations can be made:

- Part production for each part type.
- Number of pallets required to support production.
- Time in system.
- Cutting or 'tape' time.
- In-process storage time.
- Transport time.
- Station utilisation:
    - cutting.
    - material handling.
- Queue lengths (time in queues).
- Transporter utilisation:
    - moving.
    - shuttling.
    - idle.
- Transporter cycle time.
- Computer reaction time required.

Risk avoidance by simulation

The target area in this respect is the functionality of a system. By the correct application of simulation and analysis of output, the required performance can be assured against both predetermined fixed or variable criteria and also a wide range of 'what if?' tests can be applied to check out potential weak links in the system, such as transporter speed. (What if the actual transporter speed is only 80% of the manufacturer's figure?) It is thus possible to substantially reduce this risk element and hence provide improved control over the cost and timing parameters.

Current availability of simulation services

A variety of options are currently available to the prospective FMS purchaser:

- Hardware supplier.
- Independent specialist.
- In-house expertise.

Hardware supplier

It would be totally wrong and inaccurate to imply that a machine tool vendor does not work in his customer's best interests. The supplier may, however, adopt a different philosophy to determine overall priorities and cost benefits. When considering delegation of system design to a machine tool builder, several important items should be investigated:

- Capability to supply/manage total project.
- Ability to provide and explain simulation output.
- Ability to justify final design against alternatives.

The first item is obviously not relevant to this paper – except that the choice of system supplier may represent the single greatest risk.

The origin of the simulation capability provided by a machine tool builder will normally be via technical links with academic institutions, or more rarely an internal simulation capability. The former simulation source is not itself without risk:

- Third party (translation of requirements).
- Stability of supply.
- Prototype or unproven software.
- Non-accountability for errors.
- Third party: evaluation/explanation to customer.

Some of the more technically advanced machine tool builders now boast of an internal simulation capability which, apart from the question of priorities and cost benefits, may represent in the future a straightforward approach. However, if we examine the number of systems and the development cycle required by an FMS it is clearly evident that few projects can be managed concurrently. Under these circumstances it is difficult to support a team or even a single specialist due to the relative infrequency of projects. This can lead to extended input generation and validation times and also a greater incidence of errors due to slippage back down the learning curve. With time, hence experience, and assuming that the simulation specialists can be retained, these risks will gradually reduce.

With the primary motivation of most machine tool builders still being machine tool sales it is difficult to break from inbred manufacturing concepts and project organisation. Few hardware suppliers would consider utilising simulation as the primary design tool, instead relying on expertise established during a different technological age to create layouts and logistics. Any alternatives proposed by hardware suppliers would normally be restricted to that which they can both supply and make profit from. There are many examples of systems which have been designed on the assumption that the machining centre is the only machine type necessary in a manufacturing system. In many cases of course, this may be true due to the high level of inherent flexibility contained within modern machine tools.

In addition to turning centres, it is possible to communicate effectively with many other types of machine tools which have in recent years gained computerised control systems and the provision of some form of tool storage/changing device for:

- Honing.
- Gun drilling.
- External/internal grinding.
- Spark erosion.

The problem facing the prospective FMS user who has a requirement for alternative machining processes is that he may be forced to pursue the concept of 'islands of automation' with each island containing a separate major processing type. The danger in this case is that each island remains just that, with little hope of future integration due to different philosophies adopted by the individual island builders:

- Part handling:
    - size of pallet.
    - type of pallet.
    - height of working area.
    - work handling devices.
    - method of fixturisation.
    - component orientation.
    - machine, fixure, pallet error compensation.
- Communications (selection of host):
    - level of DNC.
    - communications protocols.
    - type and size of data.
    - frequency of communications.
    - response time.
    - voltage.
- Control:
    - scheduling of work.
    - scheduling and timing of transport system.
    - maximum size of system (larger than sum of islands).

Independent specialist

In this context, the word independent means having no commercial arrangement with external bodies which might affect the end result. As in most fields there are specialists ... and there are specialists. The dividing line must be drawn between theoretical and practical specialists. It is a fairly straightforward procedure to design, on paper, an FMS with all of the required features and attributes. It is another matter completely to make it work and to apply the necessary rules and constraints which are needed in the real world.

As with any application of external consultancies it is important to have some defined objectives prior to the initial contact and, for simulation in particular, to select a consultancy with the following minimum attributes:

- Access to at least one major simulation program.
- Ability to provide training.
- Willingness to supply and explain model input code for universal simulators.
- Capability to analyse results and make recommendations for changes.
- Practical experience with all phases of FMS introduction.
- Ability to supply simulation references.
- Capability to suggest the type and scope of 'what-if?' tests to be applied to the model.
- Ability to advise on simulation alternatives.
- Awareness of the need for customer participation.

The principal advantage of the true consultant is obviously the freedom from constraints. Without any practical background this may also be a disadvantage. However, this factor apart, the external consultant is able to propose a system design optimised to accommodate the types of variables which can be identified with the customer. Freedom from hardware or software constraints enables the consultant to select the best combination of machine tools, handling system, PLC or microprocessor for the job. Yet, again, without a background of practical installation the proposed optimum may not be achievable.

Experience shows that the initial runs of a simulation program with a user-defined model serve to build the confidence of the user in the validity of his model. There is also an obvious need to have confidence in the simulation program itself. This can only really be gained by constant use and with many different system types or problems being simulated. The advantage of the independent specialist is his ability to maintain expertise and judgement by frequent and diverse hands-on experience.

In-house simulation

To have total control over a function such as simulation could be an attractive proposition when considering potential savings and total system cost, and then comparing these with the cost of setting up in-house simulation. Typically, the cost of establishing a simulation capability through the purchase or leasing of hardware and software, can normally be justified against a single FMS. However this route itself is not without pitfalls, many of which may be attributed to a lack of forethought or awareness.

Since we are considering simulation as a tool, we must also decide to what purpose it will be applied. That is, is simulation required to:

- Assist in the conceptual design of an FMS?
- Create alternative concepts from which an optimum can be selected?
- Check the validity of a system already designed?
- Check the validity of an externally supplied simulation?

Also, once the project has started, will the simulation be required for:

- Short-term updates during design/implementation phases?
- Long-terms updates, to be used as a tool when considering product, volume, mix or process changes in the future?
- Additional, new projects?

At this stage, and based upon the assumption that the in-house approach is still favourable either in isolation or parallel, we must decide on the type and quantity of human resource necessary.

Human resource for in-house simulation

All of the parameters necessary for the design of a simulation model will normally be supplied by the production engineering department and, therefore, this is the logical area in which to implement the capability. A minimum of two engineers should be fully conversant with the simulation since the risks associated with a solitary specialist within a company are obvious!

Initial training time will be in the order of one week although training costs can vary considerably; for example, £500 per week per engineer external, and £400 per day per company.

It should not be assumed however, that a fully proficient FMS analyst/designer will emerge from a single week's tuition. That will not happen, but the ground rules will have been established. The engineer will learn his most important lessons when trying to apply simulation to problems not discussed in the classroom. There is no substitute for practical application.

Where time does not permit a normal learning curve progression, an alternative approach is to customise the training course to suit the particular requirements of an individual company. This way may be more expensive, but in the short term the benefits are greater due to the creation of a sound base upon which the 'students' can build and gain confidence.

Selection of a simulation strategy

In choosing a corporate strategy for simulation it is of course advisable to examine both short- and long-term objectives. Analysis of short-term objectives alone could, depending on the FMS source, justify any of the three sources of simulation already discussed depending on the levels of risk identified. However, if the total spectrum of manufacturing activities is examined, including an extended horizon and time scale, several additional risks may then become apparent. The most important risk relates to the future of the current FMS. Obviously all identifiable future risks for an FMS should have been examined during the initial design stage simulations, but it is not always possible to identify the precise nature of future changes to the product, volumes or processing techniques against which the FMS was designed. In this instance it is essential to be able to utilise the system design model at a later date. This will usually be possible if:

- In-house simulation is used.

or

- A widely used/supported simulation program has been used by a consultant or FMS builder provided that the input coding is available to the FMS user. (This is unlikely.)

If in-house simulation is not used it is important to understand that some external consultants and machine tool builders may attempt to generate forward business from withheld information or partial retention of data. As a purchaser of a service it is essential to ensure that all related information required is obtained, including model data. This approach will allow the FMS user to obtain competitive quotations for future simulation work.

A second major risk which may be identified by longer term analysis is the creation of additional new manufacturing systems. These could be systems for different products, requiring totally different hardware, or could perhaps be extensions of the current FMS.

To summarise, the strategies which are advocated in the best interests of the purchaser are either in-house simulation or any external source provided that:

- The external source is willing and able to supply all necessary data to enable either the purchaser or an alternative external body to continue with the model development in the future.
- The external source is able to provide training and documentation to explain the model concept, limitations, etc.
- The simulation program is not exclusive to the supplier or consulting organisation.

Selecting a simulation tool

The primary concern here is selection of a simulation method for in-house application, although many of the guidelines suggested here are equally applicable to external sources especially where there is a possibility of building an internal capability from an external model base. The factors which must be taken into account when choosing an internal simulation tool are:

- Resource type
  - production engineer.
  - programmer/analyst.
  - other.
- Experience
  - of simulation.
  - with computers.
  - of FMS concepts.
- Level of responsibility
  - to suggest changes.
  - to verify externally supplied data.
  - to affect project direction.
- Cost
  - in relation to anticipated savings.
- Availability of hardware
  - computer.
  - suitable output devices.
- Timing
  - availability of sufficient time for learning curve and useful application.

A widely accepted misconception exists in the field of simulation whereby a correlation is made between complexity and accuracy of results. This is of course understandable but not necessarily correct. If we take GPSS for example as one of the leading general purpose simulators it is possible to simulate almost any conceivable system with any level of detail required. However, to reach a satisfactory level of expertise requires a relatively high degree of training which must be built upon a programming background to gain the fastest possible response in the short-term. Some strict guidelines are necessary to control a simulation capability when using such a powerful tool, otherwise there is an understandable risk of input errors which may without adequate control of standards, be difficult to diagnose. As with any programming language, it is very important to control the structure of programs and to provide as much internal documentation as possible to assist the fault-finding process.

It is worth checking on the transportability of simulation programs since some are machine dependent, others may only be available as limited versons on alternative hardware, whilst others may not be supported at all except on designated equipment.

If we are examining simulations applicable to FMS then we can examine the opposite end of the spectrum. That is, easy to use, yet still powerful simulators, such as MAST used by GEC Mechanical Handling Limited which can run on a wide range of computers. In this case the program has been designed for application by production/methods engineers and is easy to learn and apply.

Concluding remarks

To satisfy the primary object of simulation – that is, the avoidance of risk – we must obviously aim to select a credible source and to be confident that the ouput, whether graphic or otherwise, can be accurately interpreted.

Generally the choice between in-house or external resource is dependent upon timing, estimated savings and the potential for future application. *It is important to note that flexible manufacturing systems are about software and control, not hardware.* For example, an effective FMS could incorporate manual workstations provided that functional control is provided by a common host. For this reason, plus hardware dependency factors, it is suggested here that the hardware supplier should not be considered as a principal simulation source, although this route may be effective when results can be verified either by independent or internal sources.

SIMULATION
AS A CIM PLANNING TOOL

R.J. Miner
formerly of Pritsker & Associates Inc., USA

A computer-integrated manufacturing (CIM) system is a complex integration of various hardware and software components. Designing such systems is often a difficult process due to the number of different components and the numerous interactions between them. Computer simulation is a proven means of describing and analysing systems consisting of many complex and interacting components. A simulation model which can be displayed graphically is also an effective communication tool, providing marketing, engineering and manufacturing with a framework within which business alternatives can be developed and evaluated. The use of simulation to design new CIM systems is discussed.

The driving force behind the increased automation of manufacturing operations is the need to stay healthy and competitive in today's quickly changing manufacturing world. Manufacturers are attempting to take advantage of current computer technology to maintain or gain a competitive edge. Using current computer technology, they consider automating their factories if substantial cost savings or increased revenues will result. To automate a factory, one must begin with a clearly defined computer-integrated manufacturing (CIM) strategy which considers all aspects of the business, from order entry to manufacturing. The design of a CIM system cannot be myopic. One must consider the entire company operation to avoid creating 'islands of automation' which will result in a suboptimal overall operation.

Integration is the key to any CIM implementation. Computer simulation is a comprehensive analysis tool which allows one to focus on individual components of a manufacturing company without overlooking the integration issues. Thus simulation can be an especially effective tool when planning for CIM.

Key aspects of CIM systems

Computer-integrated manufacturing is a method of accomplishing the entire business of manufacturing by automating and integrating as many functions as possible. It is more an evolutionary process using available computing technology, both hardware and software, than it is a particular mode of manufacturing operation. Ultimately, CIM will integrate all data processing functions, including financial accounting, purchasing, inventory, distribution, payroll, engineering and management as well as the traditional manufacturing operations. Currently few, if any, companies are integrated in this ultimate sense, but several are putting some of the key components in place.

CIM is an attempt to place management of manufacturing companies in a position to compete effectively in future markets. The job of management in the future will involve quickly identifying viable marketplaces, developing new products, producing new products in existing facilities and bringing new high-quality products to the market at low cost. To accomplish these tasks with sufficient rapidity, the future manager will need as much help as possible. CIM will need to provide these managers with:

- Flexibility in product design and product mix.
- Quick response to market demands through changes in product mix, manufacturing throughput and the utilisation of manufacturing equipment.
- Better control of all manufacturing processes.
- Reduced scrap through maximum utilisation of raw materials, equipment, personnel and inventory.
- Greater knowledge of current manufacturing operations.
- Faster throughput by maximising the use of equipment and personnel while minimising inventories, breakdowns and work stoppages for missing parts and materials.

For companies to achieve the benefits of CIM they need two items. First, management needs a clear, well-developed CIM strategy and an effective plan to execute that strategy. Second, management needs to make a commitment to an ongoing review of current computer technology to watch for, or develop themselves, advances which could accelerate the plan or require its modification. The results of acquiring and adopting computer-based technology without this type of management commitment could well be negative. There is a real possibility that the cost of choosing a series of systems which do not work or are difficult or impossible to integrate may be greater than the cost of waiting until compatible, integrated, turnkey systems are available.

A commitment to a CIM strategy will eventually result in substantial benefits. CIM will afford managers more accurate information about their operations, in both aggregate and day-to-day terms, as well as actual current status. This increased knowledge will enable them to address key problem areas quickly, reducing work-in-process (WIP), increasing utilisation and extending working hours while reducing labour costs and reducing floor space. In addition, managers will gain the ability to respond more quickly to a variety of changes from market variations to equipment failures. Production change-over costs will go down and EOQ (economic order quantity) will be reduced. Overall, CIM will result in the centralisation of information and control which will make for better management decision-making in marketing, engineering and manufacturing.

Table 1 shows some potential components of a feasible CIM system utilising currently available computing technology. A great deal of integration is necessary to link these components into a CIM system. All computing hardware must be linked together to maximise information communication and data validity, and to minimise

Table 1   Potential components of a CIM system

Business phase	Computer technology	
	Hardware	*Software*
Quotation/order placement	Mainframe computer	Automated quotation system On-line order entry system Central database
Cost estimating	Mainframe computer	MRPII Central database
Product design	Mainframe computer Minicomputer Microcomputer-based workstations	GT system CAD system MRPII Facilities design Central database
Process planning	Mainframe computer Minicomputer	CAPP system MRPII Simulation Scheduling Sequencing Tool planning Routing planning Operations planning Central database
Quality planning	Minicomputer	Inspection planning Central database
NC programming	NC, DNC, CNC machines	NC programming system

Manufacturing control	Mainframe computer
	Minicomputer
	Microcomputer-based
	workstations
	NC, DNC, CNC machines
	AMHS
	FMS
	ASRS
	Robots
		Factory management system
		Shopfloor control system
		MRPII
		Simulation
		Quality assurance tracking
		  system
		Maintenance/reliability
		  tracking system
		Production control system
		Part tracking system
		Labour reporting system
		Inventory status system
		Central database

information repetition. On the software side, all the software systems used in the various business phases must be integrated via a common database to allow information to pass from one phase to the next. Concurrent access to the database by multiple simultaneous users is essential. When a review is made of the necessary components and all their potential interactions it becomes clear why so few true CIM systems have been implemented. The process of developing components and interfacing them effectively is obviously very complex. It is clear that a means of describing components and their operations and interactions is necessary. Simulation modelling is an analysis tool which can help address these difficult CIM planning issues. The next section will provide suggestions and guidelines for using simulation in the CIM planning process.

Simulation

Computer simulation is a powerful analysis tool which can address many CIM planning issues. Simulation is the process of designing and developing a mathematical-logical model of a system and conducting experiments with the model on a computer. Simulation is well suited to address CIM planning issues for several reasons. First, a simulation model can represent the important physical and operational character-istics of a CIM system, including all the complex interactions which occur between the integrated components of such a system. Because these characteristics can be represented directly in the model, assumptions need not be made in the model which do not hold for the real system: a simulation model can be an accurate representation of reality.

Secondly, a simulation model behaves dynamically just like the actual CIM system. This allows the model to serve as a laboratory in which to evaluate the 'what if?' questions typically considered in a planning phase. Alternative designs can then be easily and consistently evaluated in a controlled environment.

The third feature of simulation which makes it especially well suited to CIM planning situations is its ability to measure the simulation model performance in the same terms that the actual CIM system would be measured. Order processing times, manufacturing throughput and resource requirements are examples of performance measures of concern in CIM systems.

Since the simulation model can accurately represent reality, behave dynamically and produce critical measures of performance, various CIM plans can be evaluated in terms of their ability to achieve overall CIM goals and strategies before a particular CIM implementation is installed. It is typically more cost-effective to evaluate alternative plans or implementations with a simulation model than with actual hardware and software.

To be an effective CIM planning tool, a particular simulation capability must have certain characteristics. First, it must be able to represent all the components of a CIM system. Secondly, it must be able to represent all of the plans to be evaluated. Thirdly, it must produce performance measures in a form easily understood by the management selecting the CIM plan. Fourthly, it must be easy to use or manufacturing personnel will not take the time to use it.

SLAM II, the Simulation Language for Alternative Modelling, is a simulation language which can accurately represent CIM systems and alternative implementation plans. When used in conjunction with TESS, The Extended Simulation System, all of the required CIM simulation characteristics are met. The examples in the following section illustrate the application of SLAM II and TESS to CIM planning.

CIM planning with simulation

Once a corporate CIM strategy has been developed, it will be necessary to demonstrate confidently that a particular automation plan can be implemented and operated in support of the corporate CIM strategy and overall business objectives. Simulation of various plans in light of the CIM strategy will give managers confidence in their automation plan, having minimised their risk of selecting a poor automation alternative.

Earlier in this article the requirements of simulation as a CIM planning tool were identified as:

- The ability to represent all components of a CIM system.
- The ability to represent all CIM plans to be evaluated in light of the CIM strategy.
- The ability to produce the measures of performance desired by management in an easily understood form.
- It must be easy to use.

Each of these requirements will now be considered individually. First is the ability to represent all components of a CIM system. To date, no special-purpose CIM simulation languages have been developed. The best simulation tool currently available for CIM applications is a general-purpose simulation language such as SLAM II, GPSS or SIMSCRIPT. Any of these proven languages will allow CIM systems to be represented accurately within its modelling framework. SLAM II has the advantage of including a set of generic modelling symbols which can be interfaced to FORTRAN code if additional detail is required. Fig. 1 shows a SLAM II network describing an automated manufacturing system which performs machining operations on castings. Fig. 2 is a diagram of the system represented in the SLAM II model. The castings are initially loaded onto pallets (carrying 16 parts each) and sent to one of two lathes via conveyor. Upon completion of the turning operation, the parts are transported, again by conveyor, to a wash/load area before being sent to the machining centre via a wire-guided vehicle.

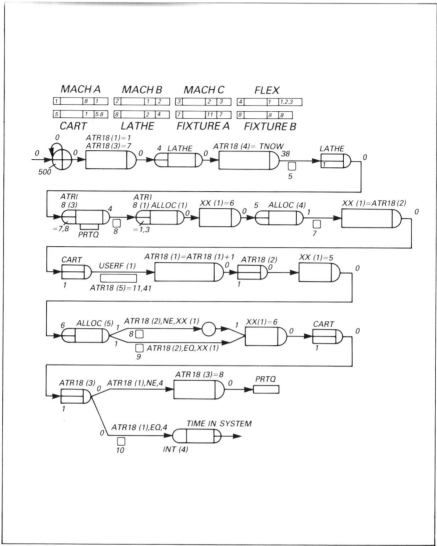

Fig. 1   SLAM II network

The machining centre consists of ten identical horizontal milling machines which can perform any one of three operations. The three operations are referred to as 10, 20 and 30. For a particular part type, the machines are set up to be either dedicated (i.e. assigned to only one operation) or flexible. Two types of fixtures (A and B) are used in this system. Fixture A is used for Operation 10 and Fixture B is used for Operations 20 and 30.

Before a set of parts is sent to a machining operation, each part in that set is attached to a fixture and sent through the wash station. When buffer space becomes available at one of the machines capable of performing the next operation, this set of parts is transported to that

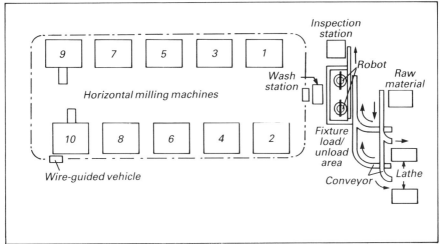

Fig. 2 Automated machining centre facility

machine's input buffer. The buffer capacity at each milling machine is one set of parts. After the parts have been machined, a wire-guided vehicle returns them to the wash/load area. The parts are then attached to a new fixture to await the next machining operation. (In the case of Operation 30, the parts are rotated 180° on the same fixture (i.e. Fixture B).) After all three machining operations have been completed, the parts are sent by conveyor to final inspection, for subsequent departure from the system. In this situation, only the operation and control of the physical manufacturing system was modelled. Although it would increase the size of the model, the operation of the software which interfaced with this type of system could also be included in the model.

The second requirement of simulation as a CIM planning tool was identified as the ability to represent all CIM plans to be evaluated. Because SLAM II is a general-purpose simulation language, it is capable of representing variations on this manufacturing situation or completely different solutions to manufacturing needs.

To evaluate each alternative CIM plan, it is recommended that the strategy outlined in Fig. 3 be followed. By starting with an aggregate model and building detail in as necessary, one can focus on the critical system components and interfaces without necessarily building a detailed simulation model of an entire CIM plan. Alternatives to the casting manufacturing situation shown earlier could be systematically evaluated using this strategy.

When SLAM II and TESS are used together, the database capabilities of TESS permit easy analysis across CIM plans. Outputs from models representing alternative CIM plans may be stored in the TESS database for later simultaneous display. The simultaneous display facilitates comparison of alternatives and selection of the best CIM plan.

Fig. 3   A general CIM
simulation modelling
strategy

Define hardware and software components
of the CIM system
↓
Define interfaces between
components
↓
Build an
aggregate-level
company-wide CIM model
↓
Model critical interfaces in detail
↓
Exercise the company-wide CIM model with
components represented at an aggregate level and
critical interfaces represented in detail
↓
Build models of critical
hardware and software
components
↓
Integrate the component models
into the company-wide CIM model,
matching inputs and outputs
of components and interfaces

The third requirement of simulation as a CIM planning tool is that it must be able to produce the measures of performance desired by management in an easily understood form. Fig. 4 contains examples of typical CIM performance measures of interest to management. In addition to these, measures of the effectiveness of hardware/software interfaces may also be desired. Any simulation capability used to analyse CIM systems must be able to provide this information in a manner usable by the simulation analyst, but easily understood by management. The outputs summarising these performance measures should be labelled in terms familiar to management, not in the language of the simulation analyst.

Hardware
Throughput of produced parts
Utilisation of hardware, personnel
WIP levels
Ability to meet deadlines
Hardware reliability information

Software
Utilisation of the database and other software
Software reliability information
Information integrity measures
Transaction processing levels
Numbers of information requests/additions/changes
Number of orders processed
Frequency of management reporting

Fig. 4   Typical CIM
performance measures

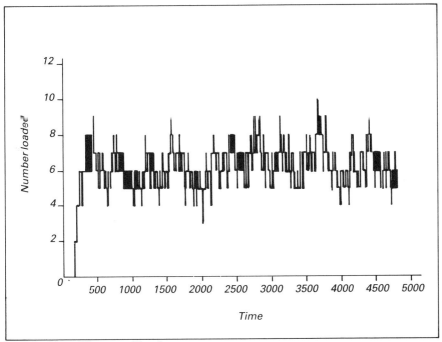

Fig. 5   Plot of fixture utilisation

Within TESS the simulation analyst has capabilities to manipulate data at a detailed level and yet have it displayed in the aggregate forms used by management. The simulation analyst defines the form in which data is to be displayed and which data should be displayed. TESS locates and displays the data. A variety of types of simulation data displays are available with the SLAM II and TESS combination. Standard columnar reports providing data at any level of detail are available as well as an extensive array of colour business graphical displays. Colour animations provide a pictoral view of the dynamic operation of the model. This comprehensive set of performance measure display types provides the simulation analyst with a great deal of flexibility to study the dynamics of a CIM model operation as well as the ability to present results to management in a particular desired form. Consider the TESS plot of fixture utilisation (from the casting model described earlier) shown in Fig. 5. The plot compares the utilisation of Fixture A with that of Fixture B. Fig. 6 shows a TESS pie chart of automated guided vehicle (AGV) utilisation for the optimal system configuration. Fig. 7 shows a histogram of throughput as a function of machine mix. Fig. 8 shows a snapshot of the animation of the operation of the optimal system configuration. The display of the animations on a colour graphics terminal shows the actual dynamic behaviour of the model. The simulation analyst selects from numerous options available to define graphical displays, including colours, location of display and labelling.

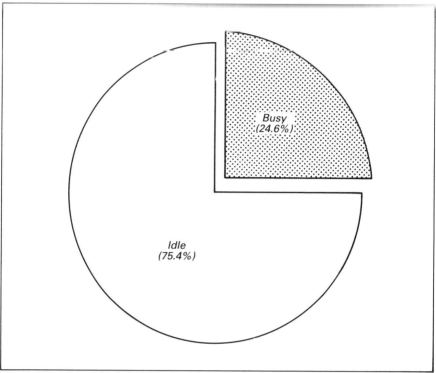

Fig. 6   Pie-chart of AGV utilisation

The fourth requirement of simulation as a CIM planning tool is that it must be easy to use. It must be easy to describe the CIM plan, exercise the simulation of the plan as desired, and get the reports of the

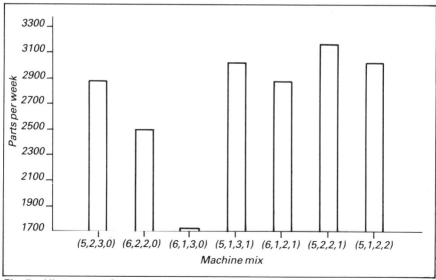

Fig. 7   Histogram of throughput comparison

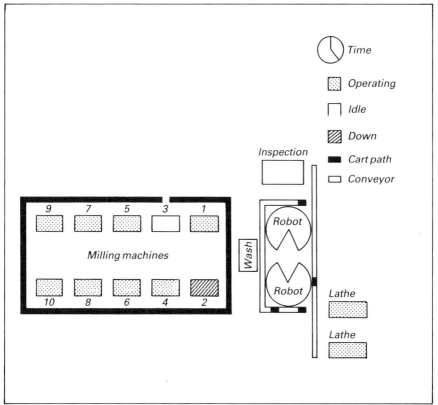

Fig. 8   Snapshot of model animation

necessary information – all in a timely manner. Any simulation tool which requires that a computer programmer codes input facilities, simulation operation and output generation will result in time-consuming CIM plan analysis. The SLAM II and TESS combination provides many ease-of-use features which enhance CIM analysis. For example, the modelling process is interactive with TESS, which allows an analyst to build graphical SLAM II models with any one of many common colour graphics terminals. In addition, the TESS language is non-procedural – that is, the analyst tells TESS what to do (PLOT UTILISATION), not how to do it. SLAM II and TESS relieve the simulation analyst of many computer programming functions by providing general capabilities interfaced with a non-procedural user interface language. In addition, SLAM II and TESS are operational on a variety of computers with a variety of graphics hardware devices. Thus this software is easily accessible to and usable by simulation analysts.

As this section has described, effective use of simulation as a CIM planning tool requires comprehensive simulation and simulation support software. Time spent in evaluating and selecting the best tools before beginning an active CIM planning phase will result in substantial time and cost savings over the entire CIM planning phase. In addition, the simulation model of the CIM plan selected for implementation will provide a basis for comparing future automation alternatives.

CIM simulation is a tool which will help marketing, engineering and manufacturing managers meet the challenges of the future. It will allow them to develop systems which provide product design and product mix flexibility. It will give them the capability to evaluate potential responses to changes in market demands. It will give them a means of better understanding the CIM system, which will improve their ability to operate and control the manufacturing process. It will also provide a tool for quantitatively measuring the potential improvements in CIM system throughput and utilisations of raw materials, equipment, personnel and inventory, which will improve their decision-making abilities. In short, modelling CIM with simulation will help place the management of forward-thinking manufacturing companies in a strong future competitive position.

Concluding remarks

While the potential benefits of a completely automated and fully integrated manufacturing system are many, few manufacturers have yet achieved CIM implementations. Reasons include the lack of a clearly defined overall CIM strategy and of a means for measuring specific implementations against the strategy.

Simulation modelling is a tool capable of dealing with the numerous complex interactions inherent in a truly computer-integrated manufacturing system. The use of simulation throughout CIM planning will reduce the risk of a less than successful CIM implementation.

References

[1] Burgam, P. 1983. Marrying MRP and CIM. *CAD/CAM Technology*, Winter 1983: 25-27.
[2] Drozda, T.J. 1983. The challenge of manufacturing management. *Manufacturing Engineering*, September 1983: 87-88.
[3] Drozda, T.J. 1983. Using computers for better manufacturing control. *CAD/CAM Technology*, Fall 1983: 15-17.
[4] Fox, K.A. 1984. MRP-II providing a natural 'hub' for computer-integrated manufacturing system. *Industrial Engineering*, October 1984: 44-50.
[5] Hegland, D.E. 1984. The challenge of computer-integrated manufacturing. *Production Engineering*, October 1984: 52-60.
[6] Searls, D.B. 1984. The automated factory: A strategy for the 80's and 90's. In, *Synergy 84 Conf. Proc.*, pp. 190-193.
[7] Stauffer, R.N. 1984. General Electric's CIM system automates entire business cycle. *CIM Technology*, Winter 1984: 20-21.

[8] Waldman, H. 1983. CIM at GE's factory of the future, *CAD/CAM Technology*, Fall 1983: 9-12.

[9] Wilson, J.R. and Miner, R.J. 1984. Simulation: A cost effective tool for the design of FMS. In, *Synergy 84 Conf. Proc.*, pp. 113-116.

[10] Wortman, D.B. and Miner, R.J. 1984. Designing flexible manufacturing systems using simulation. In, *Autofact 6 Conf. Proc.*, pp. 24, 29-24, 56.

3

Using Simulation in the Design of Machine and Robot Cells

This section describes simulation applications associated with machine and robot cell design. The papers show that the use of simulation at this detailed level can lead to optimum rules for robot and machine cells. The results from these detailed sub-models may in fact be modules or components of more complex manufacturing system simulations.

SIMULATION OF A ROBOTIC WELDING CELL FOR SMALL-BATCH PRODUCTION

R.W. Hawthorn, P.S. Monckton, R. Jones
Wolverhampton Polytechnic, UK

A Teaching Company project was set up to investigate the use of a three-station robotic cell to weld small batches of components. Analysis of likely products revealed the necessity to use up to 50 different components to obtain a fully utilised cell. The variation in product mix which may be presented to the robot is obviously very wide. In order to examine the way in which the cell may be used it was decided to simulate its operation. Results of using an interactive computer graphics package to examine the influence of batch size, fixture change time and component order upon efficiency of the cell are presented.

The use of robots to automate the welding process in mass production industries is well known, but their use in batch production situations has been limited. When used in batch production the robot is normally programmed to weld a complete batch of components before moving on to the next weld configuration. However, when the batch sizes are small it becomes difficult to keep the robot working continuously because of the additional jigging, fixturing and operator time required. This article examines an industrial situation where small batches of components are normal and considers the use of a more flexible arrangement which allows a number of different components and weld arrangements to be serviced consecutively by the robot. In order to optimise this situation the number of work stations, product types, mixes and quantities need to be carefully considered.

Since the capital cost and running costs of this welding cell will be high, it was decided to model the cell mathematically and employ a simulation technique to optimise the proposed working arrangement. The value of simulation to manufacturing situations has been demonstrated by a number of workers[1-3], and in general manual feeding of machining centres has been considered. The operation of flexible manufacturing systems (FMS) has also been examined, but in the few instances where a robot has been included in the simulation it has generally been concerned with machine loading[3]. The use of robots in a flexible welding arrangement has received little attention and the object of this work has been to consider how a cell might be operated to provide the most efficient working arrangement.

The system required a large degree of flexibility, since single-figure batch sizes are common and a range of products (safes) in varying sizes are made. In order to obtain full utilisation of the robot cell for the product range considered it has been estimated that as many as 50 different components will have to be processed.

With the combination of small batch sizes and the large number of different components, component and fixture change operations will be frequent. In order to minimise the effect of these operations on cell efficiency, a three-station system has been selected enabling a fixture to be changed at one station while the robot welds at the remaining two. The cell layout finally selected is shown in Fig. 1 and consists of the robot, trunnion and two-axis manipulator.

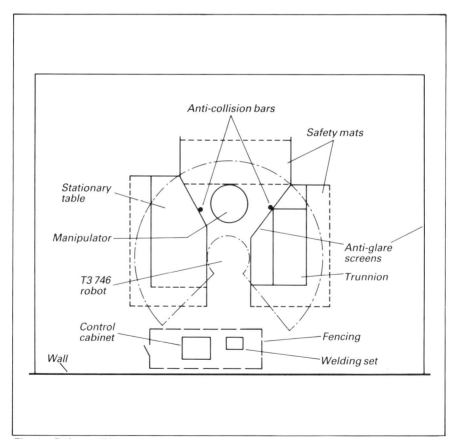

Fig. 1   Robot cell layout

Computer simulation

There are a number of computer-based simulation systems commercially available. The majority have been developed in the past five years, to simulate flexible manufacturing systems. These packages can be conveniently divided into those with and those without graphical presentation.

For this application the most efficient method of communicating ideas was via a simulation package containing graphics. Of the few systems with a graphical output, the SEE-WHY system developed by Istel Ltd was selected.

The hardware in the system consists of a Cromemco Z-2H microcomputer linked to an Intelligent Systems Corporation 8051 VDU screen. The SEE-WHY software, which consists of a set of FORTRAN callable subroutines is contained on a floppy disc in the Cromemco microcomputer. A small amount of software, approximately 8K, is housed in the 8080 processor of the VDU. This enables messages in ASCII printable characters to be converted to a form suitable for driving the screen.

The cell simulation program is written in FORTRAN IV and calls the appropriate SEE-WHY routines as required to maintain the simulation data and update the screen. The SEE-WHY system is based on discrete-event simulation, which involves the model passing forward in time from event to event, updating the status of the simulation according to the operating rules defined in the program.

Cell analysis

Model formation and validation

The cell to be simulated consists of three stations serviced by a robot. The components arrive in batches, and it was assumed that all the components would be available for welding at the beginning of each shift. This is not unreasonable since, for economic operation, the machine and press shops have to produce batches considerably in excess of those used in the welding operations.

The cell must be operated such that the robot can only service two stations simultaneously. The third station, if available for welding, will only be used when a fixture change is required at one of the two operating stations. The main reason for this is to minimise the influence of the operator on the efficiency of the cell operation. If all three stations could be serviced by the operator, it is possible that a situation could occur where two stations would be waiting for attention. The choice of station to be serviced would then be the decision of the operator, and this decision may not result in the optimum use of the cell.

One of the useful features of SEE-WHY is the ease of program debugging. This is achieved through the comprehensive error messages given and by the ability to observe the operation of the model on the screen. To ensure that the correct times for each activity are being used, the model may be run in step mode. This allows the program to be stepped from event to event, enabling the start and finish times of each event to be recorded and hence the activity time to be determined. The method used for obtaining these activity times will be discussed in the following section.

Data collection

In this particular case, the activity times have been defined as:

- Weld time.
- Load time.
- Unload time.
- Fixture change time.

The last three activities are performed manually and hence it is difficult to predict accurately the time required to perform each task. This is further complicated because the tasks cannot be directly related to present activities since different techniques of loading and unloading new fixtures are being used in the cell. The times used in the simulation program for these operations are therefore best estimates based on discussions with work study personnel. Initial trials with the cell at the manufacturers have indicated that the load and unload times of three minutes for each activity and for each component used in the model are not unreasonable.

It is easier to estimate the weld time for the components since this activity is to be automated. Following the technique described by the Welding Institute[4], the estimated weld cycle time for each component was determined. This time varied from 4 to 24 minutes for the range of components considered.

Design of experiments

Simple product mixes were initially used to evaluate the simulation model. These mixes consisted of queues of batches of equal size at each workstation. All the components in the mix had identical weld times and the fixture change times were also identical.

Because of the high capital cost involved it was decided initially to use robot utilisation as the measure of cell efficiency. In the first place the effect of batch size on efficiency was determined. This involved varying the batch size from 1-20.

The effect of weld time and fixture change time on the cell operation was then examined. The weld time of all the components was varied in the range 5-30 minutes and the fixture change time in the range 20-60 minutes. The results of these experiments are shown in Figs. 2-4. From these results a critical batch size can be determined. This is the batch size at which the robot utilisation becomes constant. Fig. 5 illustrates the dependency of critical batch size on fixture change time and weld time.

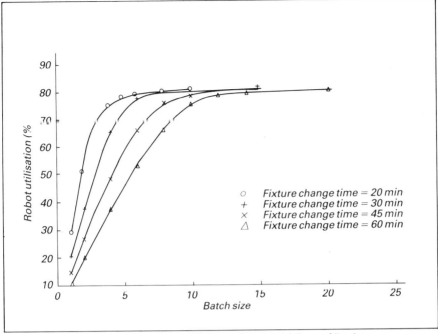

Fig. 2   Effect of batch size on cell efficiency for a weld time of 5 min

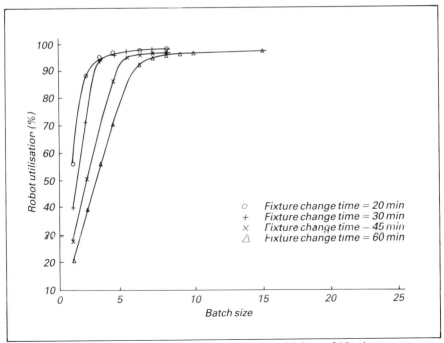

Fig. 3   Effect of batch size on cell efficiency for a weld time of 10 min

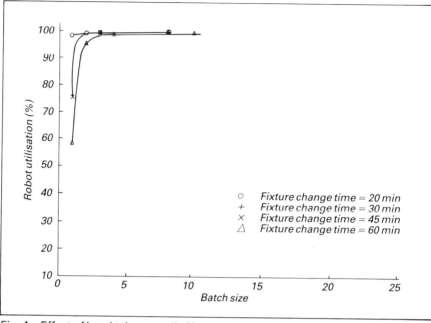

Fig. 4   Effect of batch size on cell efficiency for a weld time of 30 min

These results show that, for the weld time and fixture change time anticipated in this instance, the critical batch size is in excess of that used in normal production. As explained later, it was necessary to develop some rules for sequencing the batches through the cell in order

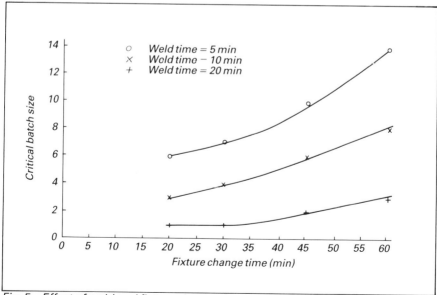

Fig. 5   Effect of weld and fixture change time on critical batch size

Table 1   Effect of varying the number of batches at each station

Number of batches at workstations			Cycle time (min)
1	2	3	
2*	5	5	436.0
2	5*	5	448.0
2	5	5*	448.0
3*	5	4	426.0
3	5*	4	448.0
3	5	4*	412.0
3*	3	3	378.0

Number of components in each batch = 3
Weld time of each component = 10min
Fixture change time = 30min
Shortest weld cycle time = 360min
* Workstation not initially serviced by the robot

to gain optimum efficiency from the system. These rules were obtained by examining the effect on the cell efficiency of the following parameters:

- The variation of the number of batches at the workstations while the other parameters remained constant.
- The sequence of various batch sizes within a product mix while the other parameters remained constant.
- The variation of weld time of components in different batches while the other parameters remained constant.

The results obtained from these experiments are shown in Tables 1–4:

Table 2   Effect of varying batch size

Batch size order at workstations			Cycle time (min)
1	2	3	
1,2,4,5*	1,2,4,5	1,2,4,5	434.0
1,2,4,5*	1,2,4,5	5,4,2,1	396.0
1,2,4,5*	5,4,2,1	5,4,2,1	416.0
5,4,2,1*	5,4,2,1	5,4,2,1	434.0
1,2,4,5	1,2,4,5	5,4,2,1*	396.0
1,1,5,5	1,1,5,5	5,5,1,1*	428.0
1,1,5,5*	1,1,5,5	1,1,5,5	466.0
2,3,3,4	2,3,3,4	4,3,3,2*	390.0
2,3,3,4	2,3,3,4	2,3,3,4*	402.0
1,3,3,5	1,3,3,5	5,3,3,1*	372.0
1,3,3,5	1,3,3,5	1,3,3,5*	418.0

Number of components = 12 at each station
Number of batches = 4 at each station
Weld time of each component = 10min
Fixture change time = 30min
* Workstation not initially serviced by the robot

Table 3   Effect of weld time

Weld time order of components at stations (min)			Total weld time at stations (min)			Stations initially serviced by robot	Cycle time (min)
1	2	3	1	2	3		
5,5,5,5	10,10,10,10	30,30,30,30	60	120	360	2 & 3	701.0
						1 & 3	696.0
						1 & 2	696.0
30,10,5,5	30,10,10,5	30,30,10,5	150	165	225	2 & 3	775.0
						1 & 3	775.0
						1 & 2	745.0
5,5,10,30	5,10,10,30	30,30,10,5	150	165	225	2 & 3	786.0
						1 & 3	786.0
						1 & 2	669.0

Number of batches at each station = 4
Batch size = 3
Fixture change time = 60min

The basic rules developed from the cell operation using simple batch mixes were then applied to realistic mixes. These mixes were established from an examination of recent schedules of safe production. In order to ensure that the rules gave optimum cycle times, a method of relating the expected cycle time to a product mix parameter was required. This relationship was determined from the results on

Table 4   Batch size order

Batch size order at workstations			Stations initially serviced by robot	Cycle time (min)
1	2	3		
1,3,6,6	1,3,3,6	6,3,1,1	2 & 3	618.0
			1 & 3	618.0
			1 & 2	677.0
6,6,3,1	6,3,3,1	6,3,1,1	2 & 3	647.0
			1 & 3	647.0
			1 & 2	647.0
1,3,6,6	1,3,3,6	1,1,3,6	2 & 3	642.0
			1 & 3	642.0
			1 & 2	642.0

Each batch contains 30min of welding

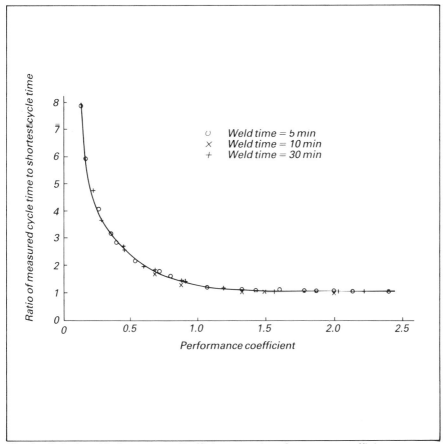

Fig. 6   Relationship between cell efficiency and performance coefficient

simple mixes as shown in Fig. 6. The product mix parameter is the performance coefficient, which relates the amount of welding in a batch to the total amount of fixture change time:

$$\text{Performance coefficient} = \frac{\text{shortest cycle time}}{\text{total fixture change time}}$$

The influence of this coefficient on cycle time was obtained by examining its effect on the ratio of measured cycle time to shortest cycle time. This ratio was used so that a standardised curve could be obtained and hence applied to any mix. The measured cycle time was obtained from the simulation. The shortest cycle time was obtained by summing the welding times of all the components assuming that the load plus unload times were less than the weld times. If for a particular component the load plus unload time was greater than the weld time, then that value would replace the corresponding weld time in the calculation.

Table 5   Cycle time optimisation for product mix A

Rules for ordering batches at workstations			Cycle time (min)	Number of batches at workstations 1 2 3	Number of components at workstations 1 2 3	Workstations initially serviced	Ratio of measured to shortest cycle time
1	2	3					
Start with min. weld quantity batches	Start with min. weld quantity batches	Start with max. weld quantity batches	601.2 611.5 595.9	4 4 4	13 8 15	2 & 3 1 & 3 1 & 2	1.639 1.667 1.626
Start with min. batch size	Start with min. batch size	Start with max. batch size	610.8 603.0 603.2	4 4 4	12 12 12	2 & 3 1 & 3 1 & 2	1.667 1.645 1.645
			622.0 627.0 630.7	4 4 4	11 10 15	2 & 3 1 & 3 1 & 2	1.695 1.709 1.721
			614.0 635.8 622.7	4 4 4	9 12 15	2 & 3 1 & 3 1 & 2	1.675 1.733 1.698
			652.0 637.0 644.8	4 4 4	9 12 15	2 & 3 1 & 3 1 & 2	1.779 1.736 1.757
Random order	Random order	Random order	678.0 613.3 669.4	4 4 4	15 8 13	2 & 3 1 & 3 1 & 2	1.848 1.672 1.825
			630.2 621.3 624.9	4 4 4	13 8 15	2 & 3 1 & 3 1 & 2	1.718 1.695 1.704
			656.0 684.2 603.2	4 4 4	12 12 12	2 & 3 1 & 3 1 & 2	1.789 1.866 1.645
Start with min. batch size	All batches same size	Start with max. batch size	620.2 605.4 619.5	4 4 4	13 8 15	2 & 3 1 & 3 1 & 2	1.692 1.650 1.689
All batches same size	Start with min. batch size	Start with min. batch size	598.6 608.1 620.2	4 4 4	8 15 13	2 & 3 1 & 3 1 & 2	1.631 1.658 1.692
Start with max. batch size	All batches same size	Start with max. batch size	630.0 622.2 622.4	4 4 4	11 8 17	2 & 3 1 & 3 1 & 2	1.718 1.698 1.698
			617.6 609.8 610.0	4 4 4	11 8 17	2 & 3 1 & 3 1 & 2	1.684 1.664 1.664

Fixture change time = 60min
Average performance coefficient = 0.679
Shortest cycle time = 366.7min

Table 6   Cycle time optimisation for product mix B

Rules for ordering batches at workstations			Cycle time (min)	Number of batches at workstations			Number of components at workstations			Workstations initially serviced	Ratio of measured to shortest cycle time
1	2	3		1	2	3	1	2	3		
Start with min. weld quantity batches	Start with min. weld quantity batches	Start with max. weld quantity batches	906.8 906.1 893.8	5	6	6	10	10	13	2 & 3 1 & 3 1 & 2	2.519 2.519 2.448
			949.1 909.7 898.9	5	6	6	5	12	16	2 & 3 1 & 3 1 & 2	2.639 2.532 2.500
Start with min. batch size	Start with min. batch size	Start with max. batch size	911.7 901.4 903.4	5	6	6	10	10	13	2 & 3 1 & 3 1 & 2	2.538 2.506 2.513
			942.4 960.9 934.1	6	6	5	11	11	11	2 & 3 1 & 3 1 & 2	2.618 2.674 2.597
			908.9 974.5 900.6	5	6	6	11	11	11	2 & 3 1 & 3 1 & 2	2.525 2.710 2.506
Smallest batch sizes at this station	Intermediate batch sizes all at this station	Remainder batch sizes at this station	922.0 890.5 905.4	5	6	6	5	12	16	2 & 3 1 & 3 1 & 2	2.564 2.475 2.519

Fixture change time = 60min
Average performance coefficient = 0.428
Shortest cycle time = 359.6min

The results obtained from this analysis are shown in Tables 5 and 6. These include not only the results from applying these rules, but also those obtained by using random sequencing of batches. The experiments were designed such that the results could also be validated by statistical methods.

Discussion of simple mix results

Critical batch size

Fig. 5 shows that for a component with a given set of conditions a critical batch size can be determined. If a batch size below this level is used, the efficiency of the cell, as measured by robot utilisation, is reduced. This critical batch size can be reduced in theory by either increasing the weld time of the component or decreasing the fixture change time. However, in practice it is difficult to vary either of these significantly, and therefore the main practical means available for increasing the cell efficiency is to increase the batch size.

As it is not normally possible to operate with very large batches, a set of conditions needs to be identified such that for a given product mix the influence of the batch sizes below the critical level will be minimised while still achieving optimum cycle time.

Performance coefficient

Fig. 6 shows that as the performance coefficient increases the ratio of measured to shortest cycle time approaches the value of 1, thus indicating that the effects from fixture changes have been minimised. The critical value of performance coefficient, at which the ratio reaches a constant value, is approximately 1.3 and is independent of weld time. Thus, if a product mix has a value less than 1.3, the mix can be classed as critical and it is therefore important that the batches should be sequenced in the correct order to optimise cell efficiency. However, if the performance coefficient is greater than 1.3, batch sequencing will have little effect on cell efficiency since all the fixture change operations can be accommodated in the cycle.

Effect of number of batches

The results so far have concerned product mixes containing the same number of batches at each work station. This seems to be a reasonable approach since three stations must be available for use during the welding of the whole product mix in order to minimise the effect of fixture changes on the measured cycle time. To verify this assumption a series of experiments was performed in which the number of batches was varied at the three stations but with the total number of batches remaining constant. The results are summarised in Table 1 and indicate, as expected, that the optimum cycle time is obtained when all the stations have the same number of batches.

Effect of batch size

Table 2 shows the effects of varying the size of the batches within a product mix. These results show that the optimum cycle time is obtained if the batches are sequenced from the minimum size to the maximum size at two of the stations and maximum to minimum at the third station. It should be noted, however, that in the sequencing of the batch sizes the weld quantity in a batch is also being ordered in the same manner since all the components have identical weld times. Therefore in the following section the effect of varying weld quantity and batch size independently of each other will be explored.

Effect of weld quantity

The initial simulation tests were performed when the weld time of all the components was identical. However, it seems reasonable to assume that the sequencing of batches with varying weld times will affect the efficiency of the cell operation. An examination of Table 3 shows that the cycle time is not greatly influenced by the total quantity of welding when all the components at each station have the same weld time. It does not appear to matter which stations were chosen to be initially serviced by the robot; the cycle time is always around 700 minutes.

If, however, the weld time of the components in the batches at a particular station is varied, then the order in which the batches are processed through that station affects the efficiency of the cell. The results in Table 4 indicate that the optimum cycle time is obtained if two stations contain batches in the order minimum to maximum weld quantity and the third station contains batches in the order maximum to minimum weld quantity. The robot must initially service the two stations that have the batches in the order minimum to maximum weld quantity. The reason for this observation is that while the initial batches are being welded a fixture change does not occur at the third station. This means that as soon as the first batch is completed the third station can be used immediately. It is advantageous to have the batch containing the smallest quantity of welding first in the queue of the work stations initially serviced by the robot. In the case of the third station, and if all the stations have the same number of batches, the last batch to be welded will be the last at this station. It is therefore advantageous to have the smallest weld quantity batch in this position, especially if it is a critical batch, since it is not associated with a fixture change.

In the previous section it was noted that the optimum cycle time was obtained when the batch sizes were in the order minimum to maximum at two of the stations and maximum to minimum at the third station. This results in the weld quantity being in the same order, since all the components had the same weld time, thus agreeing with the above observations. In order to determine the effect of batch size independently of batch weld quantity, an experiment was conducted in which a series of batches of varying sizes but with identical total weld quantity were examined. The results in Table 4 show that again the optimum cycle time is obtained when the components are in the order minimum to maximum batch size at two stations and maximum to minimum at the third station. However, in this case the robot must initially service one of the stations containing the batches in the size order minimum to maximum followed by the station with the batches in the order maximum to minimum. This is slightly different to the effect of weld quantity. With realistic mixes, where both the batch size and the weld quantity in batches vary, the relative importance of batch size and weld quantity order needs to be identified.

Rules for cell operation

The use of simple mixes in the simulation model has enabled a number of rules to be identified in order to optimise the operation of the robot welding cell. These rules can be summarised as follows:

- The three stations should contain the same number of batches.

● The order of the batches at the stations should be either:
  – minimum to maximum batch size at two stations and maximum to minimum batch size at the third

or:
  – minimum to maximum weld quantity batches at two stations and maximum to minimum weld quantity at the third.

● If the batches are ordered according to their weld quantity, then the robot should initially service the two minimum to maximum stations. If, however, the batches are ordered according to batch size, then the robot must initially service one of the two minimum to maximum stations and the maximum to minimum station.

Discussion of realistic mix results

Two typical mixes were selected to examine the rules identified in the previous section. These mixes consisted almost entirely of batches below the critical level.

The initial aims of the trials were to determine the relative importance of sequencing the batches according to size or weld quantity. The results in Tables 5 and 6 indicate that in both cases the quickest cycle time is obtained when the components are sequenced according to their weld quantity.

The second aim of the trials was to ensure that the applied rules resulted in the optimum or near optimum cycle time. Table 5 shows the other sequences that were used and the distribution of the components at the three work stations. A statistical evaluation, the t-test, was then applied to the results to check their significance. This indicated that the cell time obtained when the rules were applied to the product mix had a confidence level of 99.5%. It can therefore be concluded that these rules will result in near optimum cycle times when applied to realistic mixes.

A point worth noting is that the rules also apply to realistic mixes where the number of batches at each station is not identical, as indicated by Table 6. This table also shows that the rules apply whether or not the numbers of components at each station are similar or widely different. A slightly shorter cycle time is obtained if the numbers of components at the three stations are similar.

Finally, for each product mix an average performance coefficient can be identified. From Fig. 6 the expected ratio of measured to shortest cycle time can be obtained. The value of the ratio for product mix A is 1.7 and for product mix B is 2.6. If these values are compared with those in Tables 5 and 6 it can be seen that in general the average ratio is equal to the value suggested by Fig. 6. However, there are occasions when the ratio obtained from certain batch orders was considerably higher than that predicted by Fig. 6. It is therefore important to use these rules to sequence the components to eliminate the possibility of an exceptionally long cycle time being obtained. In general, when the rules were applied to the mixes a lower than expected ratio was obtained.

Concluding remarks

A set of rules for optimising the operation of a three-station robot welding cell has been identified. These rules apply to the situation where all the fixture change operations are identical in time. This is not unreasonable for this application, since each component will be associated with a different fixture. All the fixtures will be designed so that they can be used at any station, and therefore the methods of loading and unloading the fixtures will be similar.

Acknowledgements

The authors would like to thank Wolverhampton Polytechnic and Chubb & Sons Lock and Safe Company for their cooperation and use of their facilities, and also the SERC/DoI Teaching Company Scheme for making available the funds for this work.

The authors would also like to record their particular appreciation to Paul Hutton and Les Briggs of Chubb & Sons for their valuable contributions on the practical aspects of safe manufacture.

References

[1] Church, J. 1978. *Simulation for the Computer Novice (Manufacturing Systems)*. IEEE December 1978: 355-359.
[2] Phillips, D.T. et al. 1979. GEMS: A general manufacturing simulator. *Computer and Ind. Eng.*, 3(3): 225-233.
[3] Rathmill, K. 1980. Computer simulation of the factory of the future. *Factory Automation*, State of the Art Report, pp. 177-205.
[4] Weston, J. 1979. Welding Institute Report, 103.

MODELLING OF A CONTROLLER FOR A FLEXIBLE MANUFACTURING CELL

T.S. Chan and H.A. Pak
University of Strathclyde, UK

In a flexible manufacturing cell (FMC) the supervisor controller plays a central role in coordinating the interaction between various manufacturing, transportation and inspection machines. A systematic approach to the design of the control software for the cell supervisor is presented. This approach is based upon computer simulation of the FMC and its interaction with the cell controller, which is also modelled as a separate entity. The parameters of the models of the FMC and the controller may be altered independently and their effects may be noted prior to the implementation of the control software.

The role of a supervisory computer in a flexible manufacturing cell (FMC) is to control the interaction among the cell constituents, which include machines, parts and material handling systems characterised by their high capital cost. It is, therefore, generally accepted that a great deal of detailed analysis should be carried during the cell design stage for the selection of machinery and computer hardware and software. In order to carry out this task the user may have access to various available simulation programs which help provide an insight into the operation of the cell.

In this paper emphasis is placed on the role of the supervisory cell controller software for effective operation of the FMC. A simulation program has been developed which performs step-by-step simulation of the cell behaviour under the control action of alternative machine and transporter scheduling rules. The program consists of three sections: a model builder, an activity cycle generator and a cell controller simulator. The program caters for different types of FMC layout, such as conveyor belt systems and automated guided vehicle (AGV) systems. Several simulations have been performed to show how the system behaviour is greatly affected by some of the control rules.

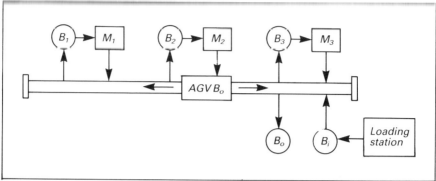

Fig. 1   FMC with AGV and local input buffer

FMC layout

In contrast to a transfer line where all parts follow the same sequence of operations, the material handling system in an FMC should permit the parts to follow a variety of different routings. Flexible routings[1] can be achieved by either providing separate paths between each pair of machines where part movement might occur, or using a common material handling device through which all parts pass and which connects all machines (e.g. a loop conveyor passing all machines or an AGV).

Within an FMC it is necessary to plan for the storage space. Two basic alternatives can be used to provide storage: local storage at machines (Fig. 1) or a common storage for the system which is accessible by all machines. There are a number of ways suggested for providing the common storage. One method of achieving flexible routing is to install a loop conveyor which is used to move parts between machines, where the parts in the loop can be considered as being in the common storage; that is, the size of the common storage is the number of spaces for parts in the loop conveyor (Fig. 2).

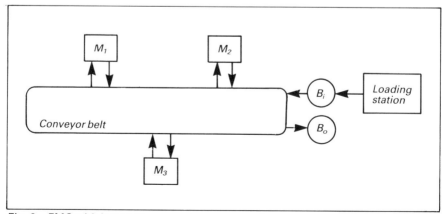

Fig. 2   FMC with loop conveyor as a common storage

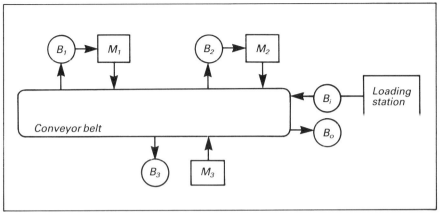

Fig. 3   FMC with local input buffer and loop conveyor as a common storage

Some systems use both common and local storage. For example, in loop conveyor systems a small number of storage spaces may be provided in front of each machine. The purpose of the local storage space is to reduce the time the machine is idle waiting for the conveyor to deliver the next part (Fig. 3).

Alternative control rules

On account of the complexity of FMCs, the potential diversity of part routing and the variability in operation times, it is necessary to give careful consideration to the control of the system at five levels as follows:

- Pre-release planning.
- Release or input control.
- Operational control.
- Loading control.
- Transportation control.

Pre-release

At the pre-release phase the parts which are to be manufactured by the system are decided, constraints on the operation sequence identified, and operation durations estimated.

Release/input control

The purpose of input control is to determine the sequence and timing of the release of jobs to the system. In this paper, three different rules are considered as follows:

- Input rate = output rate.
- Input a certain number of parts to the main input buffer periodically.
- Input a part periodically provided that there is an empty space in the main input buffer (where the part type is randomly selected).

Operational control

At the operational control level the movement of parts between machines must be ensured. If a number of alternatives exist, different rules may be chosen:

- The machine with highest priority – this may be applied to the machine which is operated at lowest manufacturing cost.
- The machine having a local input buffer of the shortest queue length.
- The machine having a local input buffer of the shortest waiting time.

Loading control

At the loading control level different rules for loading a part on a number arc considered:

- FIFO (first in, first out) in the local input buffer (Fig. 1 or Fig. 3).
- Assign a loading priority factor to each part type, so that the part type with the highest priority factor is always loaded on the machine first. This rule is applicable for a sudden demand of certain part types.
- In the case of non-universal buffers only, if a machine M_1 is blocked due to unavailability of space in the local input buffer of the next machine M_2 where the finished part type from M_1 should go for its next operation, a signal which represents this particular part type will be sent to machine M_2 so that it will machine this particular part type immediately after it has completed the present operation in order to provide an empty space in the corresponding local input buffer and eventually free the blockage of M_1.

Transportation control

In the case of a system having a limited transport facility, certain rules are applied to the components for using the transport devices. Priority is given to the component which:

- Goes for its next operation to the station where the shortest queue length exists – this may help provide a balanced workload in the system.
- Goes to the station where the shortest waiting time exists – this may help to reduce some unnecessary idle time on some machines.
- Goes to the station where its machining time is the shortest.
- Goes to the station which has been assigned the highest priority factor – this may apply to an expensive machine such that its idle time is required to be kept at a minimum.

FMC simulation program

This paper considers only one category of FMC simulator: the detailed discrete simlator[2], which performs step-by-step computation of the system behaviour. Because every event is considered in as much detail as required, it is possible to build into a model all the decision-making logic that the final system will use. This enables much more realistic predictions to be made about a system's performance.

A simulation program has been developed which consists of three main sections:

- Model builder.
- Simulation of machine process time, machine failure time, machine repair time, transportation time and all other activities.
- Simulation of the system controller.

Model builder

This requires the input data to describe the entire system, such as the number of machines, number of transporters, number of buffers, capacity of buffers, number of part mix, number of operations, routings, control rules, machining time, machine failure rate, machine repair rate, etc.

Simulation of activity time

The basis of this simulation is that the finishing time of each activity currently active is compared with the current value of a time counter. The time counter is a counter which will update one unit for every simulated time step, say one second or one minute; the time scale is a function of the system complexity and the level of detail involved. If the appropriate finishing times agree with the counter's value, the appropriate activities are stopped. The next stage is probably the most complex: it has to decide which activities may be started. Once the new activities, if any, have been allocated the appropriate times, these together with those already active are considered, and the cycle is repeated. As the performance of the system is simulated step by step, a pseudo-function generator can be used for machine breakdown simulation, and analysis of the transient response of a system is also possible.

Simulation of controller

Computer simulation techniques can be applied to all aspects of manufacturing planning, particularly in the area of control software development. Because of the complexity of FMCs, the choice of applicable control strategies depends on many particular system variables. This model of the controller is used to evaluate the performance of the planned FMC under a variety of operating strategies and external influences, (e.g. machine breakdown). Results from these experiments are assessed to decide which control rule is best suited to the planned FMC, and also to meet the manufacturers' requirements.

Major functions performed by an FMC controller[3], operating as a 'real-time' system, include the following:

Workpiece control. The controller keeps track of the location of all workpieces, both in the FMC and in queues available to the FMC. The current production status and a list of remaining work are retained for each individual workpiece. The database also stores, for each part mix, the routing, including acceptable workstations with corresponding cycle times.

Material transport system control. The controller has stored in its database the current location of all transport devices and their status (e.g. idle, carrying workpiece, going to get workpiece, etc.). Through the use of stored control algorithms, monitoring of transport system status changes, and recognition of commands from other parts of the control system, the transport system control issues the appropriate commands to move workpieces along transport pathways between stations, in a manner consistent with optimal FMC operation.

Workstation control. The control system issues appropriate instructions for carrying out the required process, for work on a given workpiece, at a given workstation. Monitoring is done for completion of the process, as well as for the occurrence of an abnormal termination or stoppage. For a station with one or more fixed cycles, the controller will initiate the start of the appropriate cycle, and monitor for abnormal stoppages and for normal cycle completion.

The controller directs the transfer of workpieces between the station and the transport system, and monitors the loading and unloading of machines. Physical verification of workpiece identity may be done at the workstation, to minimise the possibility of initiation of the wrong process cycle.

NC part program control. The various part programs required are stored by the control system and transmitted to machines as required.

Output information. The most obvious results are stated as follows:

- Production rate.
- Throughput time for each component.
- Queue length statistics.
- Machine utilisation.
- Transport device utilisation.
- Current buffer status.
- Current machine and transport device working status.
- Machine failure statistics.

Traceability. One important area of software often missed is traceability. Traceability deals with recording which operations were completed by which machines and/or operations to enable backtracking, especially of faulty manufacture to the machine station where the operation was done.

Such a facility is important:

- To stop further manufacture of faulty parts.
- To identify and correct the fault on the machine.
- To identify the actual machine tool if there are several similar ones in the system.

Examples

In order to show how the performance of a cell may be affected by employing different control rules, simulations of three different systems are presented and some of their results discussed.

Control rules for alternative routes

The manufacturing system can be described as follows:

Number of part types = 3
Number of machines = 4
Number of AGVs = 1
Transportation time between each station = 1 time unit
Condition for component entry to the system: input rate = output rate
Simulated time cycle = 450 time units
Buffer capacity (universal): interstage buffer = 10
input/output buffer = 100
Initial condition: 5 components of each part type have been loaded in the main input buffer
Working sequence: part 1 $M_1 \longrightarrow M_2 / M_3 \longrightarrow M_4$
part 2 $M_3 \longrightarrow M_1 / M_2 \longrightarrow M_4$
part 3 $M_2 \longrightarrow M_1 / M_3 \longrightarrow M_4$

Table 1a Process time for each part type

Part type	M_1	M_2	M_3	M_4
1	6	14	10	1
2	15	23	8	1
3	6	4	5	1

Table 1b Simulation results for different rules of alternative route selection

IMPRI	U (M/C)	U (AGV)	No. of finished products				Average lead time		
			part 1	part 2	part 3	total	part 1	part 2	part 3
1	64.611	97.333	21	21	14	56	94.2	95.3	142.0
2	52.444	92.889	19	15	17	51	104.3	124.0	112.3
3	60.389	99.778	19	18	19	56	100.6	110.4	105.2

Table 1a   Queue length statistics for each interstage buffer

IMPRI	Record of maximum no. of components found in each buffer				Total space required for interstage buffer
	B_1	B_2	B_3	B_4	
1	7	7	5	1	20
2	4	2	10[†]	1	17
3	5	5	5	2	17

IMPRI = 1, alternative route is selected by the shortest queue length
       = 2, alternative route is selected by the shortest process time
       = 3, alternative route is selected by the shortest waiting time
   U = mean utilisation

[†] B_3 is blocked when IMPRI = 2

The process time for each part type is shown in Table 1a. Table 1b and Table 1c show the simulation results. The dynamic behaviour of the system has been plotted in Figs. 4 and 5.

The results show that rule 1 (IMPRI=1) will give the highest mean machine utilisation (Fig. 4) and shortest average lead time for part type 1 and part type 2 (Fig. 5). The highest utilisation of the transporter (Fig. 4) as well as the shortest average lead time for part type 3 (Fig. 5) can be achieved by applying rule 3. Rule 2 offers the worst mean machine and transporter utilisation, and also gives the lowest production rate.

The reason for rule 2 producing the worst performance is that M_3 offers the shortest process time in the second operation for part types 1

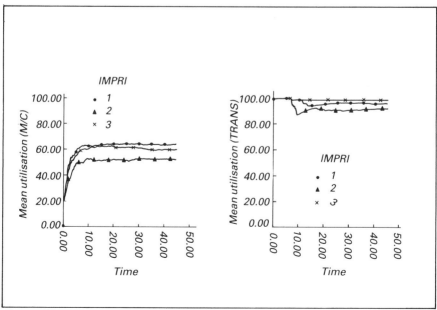

Fig. 4   Mean utilisation of machines and transporter (example 1)

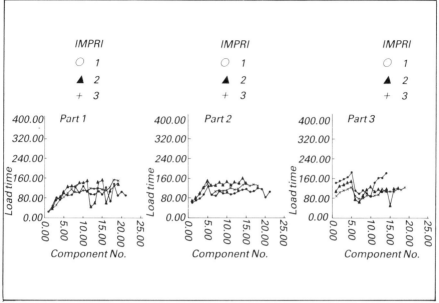

Fig. 5   Lead times for different part types (example 1)

and 3, and hence there will be an unbalanced workload between M_2 and M_3 when producing part type 1, and similarly between M_1 and M_3 when machining part type 3. In fact, M_2 will be used to machine part type 1 if and only if the local input buffer of M_3 is full. Similarly, part type 3 will not be machined by M_1 unless there is an absence of empty space in B_3. According to Table 1c, although rule 1 offers the highest mean machine utilisation, the total space required for the interstage buffers may be an uninteresting attraction.

Control rules for using transport facility

The manufacturing system is described as folllows:

Number of part types = 3
Number of machines = 4
Number of AGVs = 1
Transportation time between each station = 1 time unit
Condition for component entry to the system: input rate = output rate
Simulated time cycle – 450 time units
Buffer capacity (universal): interstage buffer = 10
                                         input/output buffer = 100

Initial condition: 5 components of each part type have been loaded in the main input buffer

Working sequence: part 1 $M_1 \longrightarrow M_2 \longrightarrow M_3 \longrightarrow M_4$
                    part 2 $M_1 \longrightarrow M_3 \longrightarrow M_4$
                    part 3 $M_2 \longrightarrow M_4$

Table 2a   Process time for each part type

Part type	M_1	M_2	M_3	M_4
1	2	10	7	5
2	2	–	7	5
3	–	10	–	5

Table 2b   Simulation results for different rules of using transporter

IDEPRI	U (M/C)	U (AGV)	No. of finished products				Average lead time		
			part 1	part 2	part 3	total	part 1	part 2	part 3
1	56.444	100.0	19	25	11	55	102.4	82.6	160.3
2	58.5	100.0	15	25	17	57	125.5	82.0	114.6
3	53.556	100.0	20	25	4	49	99.2	81.7	246.7
4	54.5	100.0	20	25	5	50	98.8	80.9	266.6

Table 2c   Queue length statistics for each interstage buffer

IDEPRI	Record of maximum no. of components found in each buffer				Total space required for interstage buffer
	B_1	B_2	B_3	B_4	
1	1	3	6	1	11
2	1	6	5	5	17
3	1	3	7	1	12
4	1	3	7	1	12

IDEPRI = 1, gives priority to delivery of the component which is going to its next operation
        station where the shortest queue length exists
      = 2, FCFS
      = 3, gives priority to delivery of the component which is going to its next operation
        station where the shortest process time exists
      = 4, gives priority to delivery of the component which is going to its next operation
        station where the shortest waiting time exists

The process time for each part type is given in Table 2a. The simulation results are shown in Table 2b and Table 2c. Figs. 6 and 7 show the dynamic responses of the system.

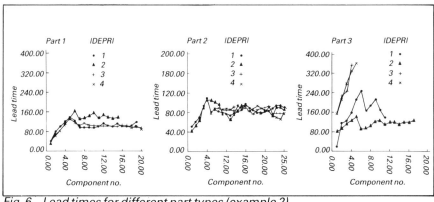

Fig. 6   Lead times for different part types (example 2)

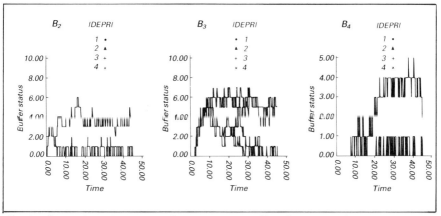

Fig. 7   Queue length statistics for interstage buffers (example 2)

According to Table 2b, the maximum number of finished products as well as the highest mean machine utilisation can be obtained by employing rule 2. On the other hand, rule 3 produces the worst results in terms of mean machine utilisation and production rate. In this simulation there is a dramatically long lead time for part type 3 if control rule 3 or 4 is applied. This is because the first operation machine for part type 3 has been allocated to M_2, which offers a longer machining time (10 time units) compared with another first operation machine, M_1 (2 time units), and hence part type 3 will receive a relatively lower priority for despatch from the main input buffer. From Fig. 6 it can be observed that the lead time for part type 3 is still in the transient state if rule 3 or 4 is employed.

Another important system parameter considered here is the size of interstage buffers. According to Table 2c, the total space required for interstage buffers is 17 if rule 2 is used, but maximum production rate and best mean machine utilisation can be achieved. Although rule 1 offers the second highest mean machine utilisation, the smallest total space required for the interstage buffers gives an attractive feature. It should be stressed here that the value of the maximum number of components found in each interstage buffer may occur only during the transient state, or within a very short period (Fig. 7). If this is the case, a lot of buffer spaces will be wasted for most of the time. By reducing the buffer capacity, the mean machine utilisation may be decreased due to the creation of bottlenecks at some stations. One can reduce the number of spaces allocated to each buffer and run the simulation again until a compromise solution is achieved.

Control rules for machine breakdown
In order to show how a system will respond when a machine has failed, a simulation of a simple system with an unreliable machine is presented.
    The system is described as follows:

Number of part types = 2
Number of machines = 4
Number of AGVs = 1
Transportation time between each station = 1 time unit
Condition for component entry to the system: launch one component of each part type to the main input buffer in every 15 time units
Simulated time cycle = 450 time units
Buffer capacity (universal): interstage buffer = 10
input/output buffer = 100

Initial condition: 2 components of each part type have been loaded in the main input buffer
Working sequence: part 1 $\quad M_1 \longrightarrow M_2 \longrightarrow M_4$
part 2 $\quad M_3 \longrightarrow M_2 \longrightarrow M_4$

Machine failure information: M_2 will breakdown randomly repair time = 50 time units

Table 3a shows the process time for each part type. The simulation results are given in Table 3b and Table 3c. Figs. 8–10 show the dynamic behaviour of the system.

Table 3a   Process time for each part type

Part type	M_1	M_2	M_3	M_4
1	8	2	–	1
2	–	2	7	1

Table 3b   Simulation results for machine failure condition

MBRKDEL	U (M/C)	U (AGV)	No. of finished products			Average lead time	
			part 1	part 2	total	part 1	part 2
1	22.833	66.667	19	18	37	93.6	104.5
2	30.5	80.444	22	21	43	77.0	82.5
3	36.111	97.333	30	30	60	31.8	38.5

Table 3c   Queue length statistics for each interstage buffer

MBRKDEL	Record of maximum no. of components found in each buffer				Total space required for interstage buffer
	B_1	B_2	B_3	B_4	
1	10[†]	1	10[†]	1	22
2	4	10[†]	4	1	19
3	1	1	2	1	5

MBRKDEL = 1, stop component delivery to failed machine
= 2, Maintain component delivery to failed machine
= 3, all machines are reliable

[†] *Buffer is blocked.*

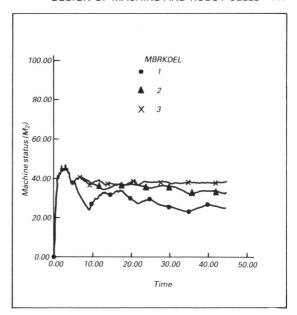

Fig. 8   Mean utilisation of machines (example 3)

The results show that the mean utilisation of machines (Fig. 8) and the transporter are greatly affected by the failure of M_2 if rule 1 (MBRKDEL=1) is applied. This is because M_1 and M_2 have been forced to stop production while M_2 is failed. For the same reason, the local input buffers B_1 and B_3 are always blocked (Fig. 10). When MBRKDEL=2, M_1 and M_3 are allowed to operate even when M_2 has failed, and hence a long queue length has built up inside the local input buffer B_2.

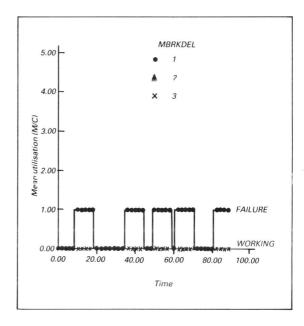

Fig. 9 M_2 operation statistics (example 3)

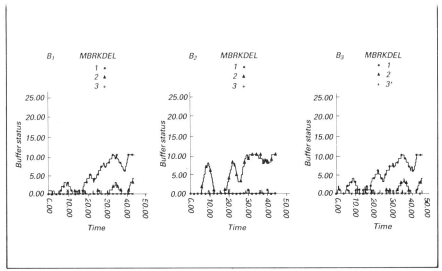

Fig. 10   Queue length statistics for interstage buffers (example 3)

Concluding remarks

Having simulated several FMS, the authors are able to make some statements about the control of a cell. In general, the best priority rule depends on the performance measure, as well as the system parameters. This study provided an insight into the modelling of an FMC controller, and the significance of employing different control rules has been reflected in the results. In order to operate an FMC efficiently, a suitable cell controller software is required. However, because of the complexity involved it is necessary to simulate the FMC controller before actually implementing the control software, which could prove to be a very expensive method of finding design errors.

References

[1] Buzacott, J.A., and Shanthikumar, J.G. 1980. Models for understanding flexible manufacturing systems. *AIIE Trans.* December 1980: 339-49.
[2] Kay, J.M. 1984. The use of modelling and simulation techniques in the design of manufacturing systems. In, *Int. Conf. on the Development of Flexible Automation Systems*, July 1984.
[3] Knight, J.A.G. 1984. Control requirements for flexible manufacturing. In, *Int. Conf. on the Development of Flexible Automation Systems*, July 1984.

MANUFACTURING CELL PERFORMANCE SIMULATION USING ECSL

B.J. Clarke and P.F. Kelly
Huddersfield Polytechnic, UK

A logical, sequential approach illustrating the formulation of a computer simulation to represent the internally variable workings of a multi-machine cell, using realistic data is given. The language used is the 'extended control and simulation language' (ECSL), in conjunction with the 'computer aided programming system' (CAPS) which generates the ECSL program. The cell under study is subjected to changes in operating parameters and the results analysed. The speed, accuracy and clarity of program construction emphasises the value of simulation in the assessment of cell performance; thus reducing the uncertainty associated with the introduction of new and expensive manufacturing facilities.

Over the past few years simulation has rapidly become one of the most widely used technical aids in operational research, administration, engineering and management sciences[1]. One reason for this is that the technique enables existing or projected systems to be experimentally tested using varying series of input data, before commitment of capital outlay. Also the development of specialised software packages with program generation facilities has contributed to the availability and relatively low cost of simulation techniques[2]. In particular, the program generator's ability to construct programs quickly and efficiently in a high-level language is seen as an important development.

In general, simulation can be described as either being of 'discrete-event' or 'continuous' type. In the former the state variables change only at a finite number of points in time (e.g. in a manufacturing cell), whilst in the latter the variables change continuously with respect to time (e.g. a model aircraft inside a wind tunnel)[3]. The logical, cyclic and progressive nature of this particular study obviously requires discrete-event simulation.

It is with these factors in mind that a simulation model has been formulated to represent a multi-manned machine gear grinding cell. Simulation runs to test the effect of varying the man/machine ratio are performed and the results analysed. The CAPS/ECSL language is chosen because of its adaptability, availability and program generator facility, thus enabling low cost programs to be constructed accurately and efficiently.

A recent tendency is to look for optimal solutions to highly simplified problems thus restricting the application of computer simulation techniques to theoretical problems[4]. In addition many manufacturing process simulations using theoretical data have been constructed[5,6]. However, this paper concerns itself with computer simulation and experimentation using real-life data obtained from an existing system at a local company.

Computer simulation model formulation

The real-life system under study comprises a five machine Reischauer gear grinding cell (Fig. 1). The machines are grouped in three lines (A, B and C), one machine occupies line A, and two machines occupy each of lines B and C. Each line is set up to grind a particular family of gears, which pass in batch sizes of 6, 25 and 50, through lines A, B and C, respectively. The cell comprises six facilities or entities: gears, machine lines A, B and C, setter and operators.

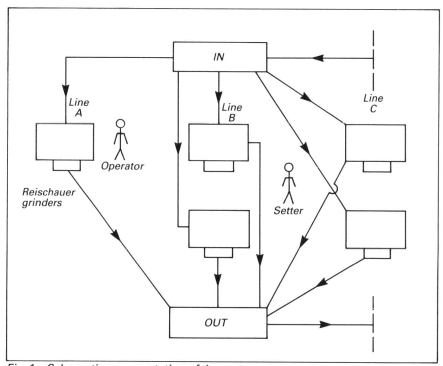

Fig. 1   Schematic representation of the system

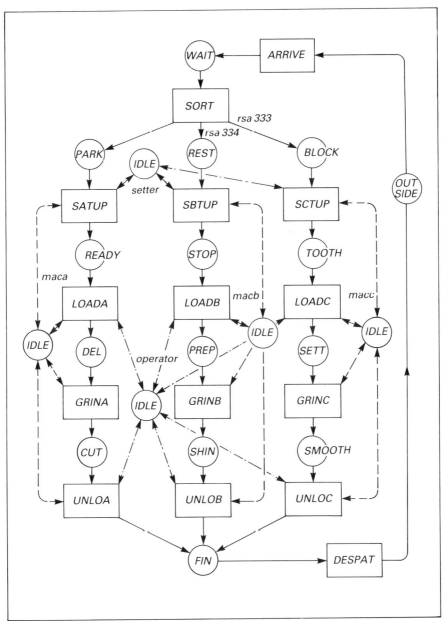

Fig. 2 Entity cycle diagram

The entity cycle diagram representation of the system is shown in Fig. 2. The cyclic interactions of these entities are verified at this stage of the model formulation. The construction of the diagram itself must be conventionally accurate. Rectangles represent activities in the system, circles represent queues, and they must alternate. A simulation 'horizon' must exist for the model. In this case gears ARRIVE from OUTSIDE and finished gears are DESPAT (despatched) OUTSIDE.

Table 1   System activity durations (minutes)

Machine line	Set-up	Grind	Load	Unload
A	80	30	2	2
B	60	10	1	1
C	60	5	1	1

Therefore the simulation model appertains only to the system enclosed within these parameters.

The cyclic representations for the machine lines A, B and C, setter and operators, have no queues. This is because there is only one state (IDLE) they can exist in when not performing activities. This one common state need not be listed when describing their cycles. However, the gear entities transfer through various states which need to be defined accurately. For example, the queue at CUT is at a totally different state from the queue at READY and must be described accordingly. Activity durations from the real-life system are given in Table 1.

Computer interface

The programming system using CAPS/ECSL follows five different stages: logic, priorities, arithmetic, recording, and initial conditions[7].

Logic

The cyclic symbolism shown in Fig. 2 is here input directly into the computer. The logic input for this particular study is fully reproduced in Table 2. The numbers shown after the entity names indicate the quantities of each entity in the system at a particular time. The computer processes and checks the logic for cyclic continuity and, when found to be correct, the next stage is entered.

Priorities

Many codes exist in ECSL to define the progress of entities through the system (e.g. FIFO, MAX, MIN, PAIR, SUIT). For this particular study the code FIFO (first-in-first-out) is the one used for all events.

Table 2   Logic input

gears 321 qoutside, aarrive, qwait, asort, qrest, rsa 334, qblock, rsa 333, qpark, asatup, *, 6, qready, aloada, qdel, agrina, qcut, aunloa, qfin, adespat, qoutside
+ qrest, asbtup, *, 25, qstop, aloadb, qprep, agrinb, qshin, aunlob, qfin
+ qblock, asctup, *, 50, qtooth, aloadc, qsett, agrinc, qsmooth, aunloc, qfin

maca (line A) 1 asatup, aloada, agrina, aunloa

macb (line B) 2 asbtup, aloadb, agrinb, aunlob
macc (line c) 2 asctup, aloadc, agrinc, aunloc

setter 1   asatup, asbtup, asctup

Opeator 1 (2, 3, 4) aloada, aloadb, aloadc, aunloa, aunlob, aunloc

The computer also lists the system activities in a relative priority order, enabling activities having highest priority to be listed accordingly.

Arithmetic

The computer lists the system activities next to which an integer number must be inserted to indicate the durations (e.g. SATUP 60, GRINB 10). Pseudorandom functions are available which generate samples from the normal, negative exponential, random and uniform distributions. The deterministic (as against stochastic) approach adopted by this study, leads to the use of real whole integer numbers.

Recording

The recording stage automatically provides a count of how many times each of the unbound activities occurs during the recording interval. This provides an indication as to whether or not the system is in equilibrium. Also a measure of the utilisation of facilities (e.g. machine lines, operators) is automatically given.

A coded integer number may be input after each of the queue names to describe the type of recording required. For example, CUT 5 means that two recordings of queue CUT are required. One being a record of the length of time an entity is delayed in the queue and the other being the length of queue recorded as a time weighted histogram. As this study is primarily concerned with production rates and utilisations over a certain time period, no information regarding the state of queues is produced.

Initial conditions

Any activities which are actually in progress may be defined here. The computer asks for a termination time for each named activity in progress. These activities use up several of the entities; for example, if all the machines are running then five gears will be utilised, and the remaining gear entities must then be assigned to various queues.

After this stage has been completed the CAPS generates the program in ECSL, ready for running.

Stratagems for simulation

The simulation runs to be carried out represent two days' production of the system. That is, two 8-hour shifts or 960 minutes using the same timescale units as Table 1 expanded. In addition, a running-in period of 100 minutes before recording takes place has been specified, in order to overcome the start-up condition problems associated with many simulation exercises. Thus the total simulation duration is $960 + 100 = 1060$ time units (minutes).

The study is directly concerned with the interaction of the man/machine ratio. Elements such as a machine line starvation, breakdowns, delays, etc. are not considered essential for the running of such a short timescale simulation. For these reasons initial conditions include the provision of two days' buffer stock, to eliminate the possibility of line starvation. Thus, sufficient gears to satisfy the demand for two days' production have been assigned to the machine line primary queues, PARK, REST and BLOCK.

In the entity cycle diagram (Fig. 2) after the activity SORT, the gear entities branch into the three machine lines (A, B and C). For simulation purposes there must be a coded instruction which channels the gears into each of the three lines. For example, rsa 333 indicates that gears will branch from SORT to queue BLOCK with a probability of 1 in 3. Codes such as this are essential to ensure the continual cyclical running of the simulation.

The set-up activities SATUP, SBTUP and SCTUP only occur when there are sufficient stocks of gears in queues PARK (6), REST (25) and BLOCK (50) respectively, as indicated by the asterisk in Table 2. For the two days' simulation run these queues have infinite buffer stocks, which remove the waiting time for stock build-up. However, if the simulation duration exceeds the two days, then this waiting time, with its attendant machine line starvation, assumes greater importance.

Results

A specimen output listing the production rates and utilisations obtained using a manning level of 1 operator is given in Table 3. The production rate through the system is given by the number of times activity DESPAT is started, in this case 290. Note also the utilisations are aggregated. For instance, if the utilisation for the operator in a two operator system is 0.5000 then each operator has a utilisation of 0.5000.

The output from all the simulation runs performed is analysed in terms of system equilibrium and re-run to validate the continuing stability of the simulation. The results themselves are presented in graphical form in Figs. 3, 4 and 5.

Table 3   Specimen output (manning level 1)

UNLOC was started	200 times
LOADA was started	48 times
UNLOA was started	14 times
LOADB was started	116 times
UNLOB was started	76 times
LOADC was started	200 times
GRINA was started	15 times
SATUP was started	7 times
SCTUP was started	5 times
GRINB was started	80 times
SBTUP was started	5 times
GRINC was started	200 times
SORT was started	289 times
DESPAT was started	290 times
Utilisation of maca	0.9643
Utilisation of macb	0.6346
Utilisation of macc	0.8207
Utilisation of operator	0.7089
Utilisation of setter	0.9782

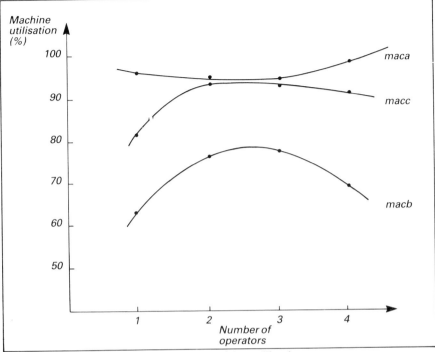

Fig. 3   Number of operators against machine utilisations

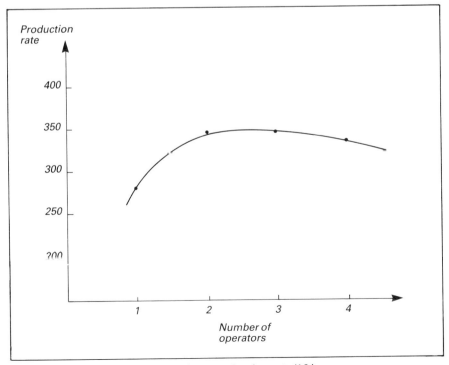

Fig. 4   Number of operators against production rate/16 hours

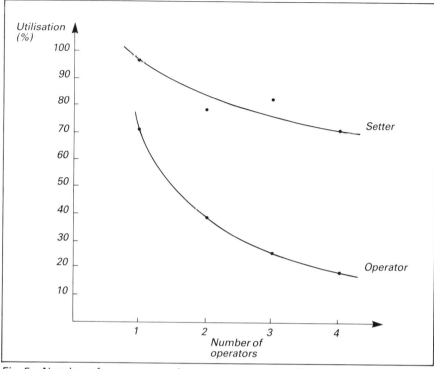

Fig. 5   Number of operators against setter/operator utilisations

Fig. 3 clearly illustrates the connection between the number of operators (manning level) and the machine line utilisations. Only in machine line A is there an increase in utilisation as the manning level increases. The utilisations for lines B and C tend to peak when the manning level reaches 2 and decrease markedly thereafter. The fact that lines B and C each contain two machines, emphasises the decrease in utilisation which occurs when the manning level increases, thus suggesting that 2 may be the most satisfactory cell manning level in this case. Noting also that the operator manning level may not be the only criterion here. The machine/setter interaction also plays a part. That is, although there is an increase in operator numbers, machines may still have to wait IDLE for a setter.

Similarly Fig. 4 shows that when the manning level exceeds 2 the production rate no longer increases. Thus in terms of pure production rates there seems to be no advantage in exceeding this level.

Fig. 5 indicates the setter and operator utilisations recorded as the number of operators increases. As expected the operator utilisation decreases markedly and is approximately equivalent to the expected ratios of 71:1, 71:2, 71:3 and 71:4. Thus providing an example of the simulation model's validity. The setter utilisation decreases slowly as the manning level increases. This graph requires a 'best-line' fit through points 2 and 3, due in the main to the lack of intermediate points, caused by the computer requirement for whole integer numbers. That is, 2 and 3 operators, not 2½ operators, etc. The highest setter utilisation occurs when the manning level is 1, but remains fairly high when the level is 2.

From the above results it seems appropriate to state that the most satisfactory manning level in terms of production and utilisation is somewhere between 1 and 2.

Concluding remarks

The simulation runs are performed on Huddersfield Polytechnic's ICL 29/60 mainframe computer, utilising the CAPS/ECSL applications package. Each separate run's request-to-spool time occupied, on average, 70 seconds. A microcomputer version of the package is now available, thus overcoming work file storage problems common to such multi-user mainframe usage.

The formulation of the model, in terms of company process routing and activity event durations, occupied by far the longest timespan in the entire study. It is essential to verify that the model is a true representation of the system under study before the simulation takes place. For an existing manufacturing system, liaison with production personnel is of the utmost importance.

The computer interface for the program construction is fast and efficient, allowing the user to concentrate on the really important simulation issues, such as validity and verified input data. The time taken to program a model for a simulation run in this study is approximately 30 minutes for a fairly experienced user. This, coupled with a delay for output of say 3 minutes, gives an indication of the speed of the compilation.

Editing facilities are available, but are not comprehensive. It is possible to change the activity durations or the simulation duration by simply editing the respective ECSL file for each simulation. However, for alterations applicable to the number of entities (operators, machines, etc.), the logic stage has to be re-entered. This is due to the fact that attributable numbers in the existing file will not correspond to altered entity numbers.

As stated earlier the most satisfactory manning level is 1 or 2 operators. In fact the company have already successfully implemented a manning level of 1 in their system, as against their previous level of 4 operators, with its attendant savings. This emphasises a priority for utilisation rather than for production.

There can be little doubt that a simulation study such as this is insular in character. The use of short timescale infinite buffer stocks allows experimentation with the man/machine ratio interaction, without the considerations of late arrivals, breakdowns, delays, etc. As an adjunct to this work, simulation runs are carried out to represent one month's production (based on the same expanded timescale). The expected machine line starvation takes place and all the utilisations recorded

throughout the period fluctuated markedly in the demand substituent time.
This illustrates that the inter-arrival rate and the branch codes probably need to be revised. However, the company does take an insular approach, assuming quite correctly that a decrease in utilisation occurs because of line starvation, due to stock build-up waiting time.

It cannot be emphasised too strongly that a series of outputs from a simulation run appertains only to a specific input.

The adaptability and power of simulation techniques using the speed and efficiency of a program generator facility such as CAPS, is shown to advantage by the fact that variable programs can be compiled so quickly, without re-writing huge amounts of routines from first principles, as is sometimes common with general purpose languages such as FORTRAN. Further simulation runs on the same system model by varying: number of machines, operators, setters, inter-arrival rates, activity durations, etc. can be compiled with little time or cost expended.

Acknowledgements

The authors would like to thank David Brown Gear Industries Ltd, Huddersfield and in particular Mr B. Carter (Principal Project Engineer) for invaluable help in amassing the data for this study. Also Mr G. Williams (Project Coordinator, Computer Academic Support Group, Huddersfield Polytechnic) for his assistance in hardware/software implementation.

References

[1] Martin, F.F. 1968. *Computer Modelling and Simulation*. John Wiley, New York.

[2] Mathewson, S.C. 1984. The application of program generator software and its extensions to discrete event simulation modelling. *Trans. AIIE*, March 1984.

[3] Spriet, J.A. and Vansteenkiste, G.C. 1982. *Computer Aided Modelling and Simulation*. Academic Press, New York.

[4] Dunham, N.R. and Kochhar, A.K. 1981. Interactive computer simulation for the evaluation of manufacturing planning and control strategies. In, *Proc. UKSC Conf. on Computer Simulation*, pp. 82-89. IPC Business Press.

[5] Ang, C.L. 1981. Computer simulation of the effects of inter-cell workload transfer on the performance of group technology manufacturing systems. University of Nottingham.

[6] Hutchinson, G.K. and Holland, J.R. 1982. The economic value of flexible automation. *J. Manufacturing Systems*, 1(2): 215.

[7] Clementson, A.T. 1978. An extended control and simulation language. In, *Proc. UKSC Conf. on Computer Simulation*. IPC Business Press.

4

Using Simulation in the Design of FMS

The papers contained in this section describe the simulation life-cycle associated with the design of FMS projects. A system design is hypothesised, and relevant product and process planning data obtained. A simulation model of the proposed FMS can then be developed and tested. This section highlights the interactive use of simulation so that numerous variants of the original system's design may be evaluated in order to select the best or most robust.

SIMULATION AS AN INTEGRAL PART OF FMS PLANNING

H.-J. Warnecke, R. Steinhilper and K.-P. Zeh
Fraunhofer Institut für Produktionstechnik und Automatisierung (IPA),
West Germany

With reference to a number of different case studies, the significance
of simulation techniques today is discussed. The range and types of
application conceivable for simulating, analysing, comparing and
evaluating FMS planning concepts, design variants and control
strategies are described.

Simulation technology has developed into an integral component of the
planning of flexible manufacturing systems. In view of the still
extensive development and standardisation work being carried out in
the field of simulation, there is hope that medium-sized companies as
manufacturers and users of flexible manufacturing systems will
increasingly reap the benefits of this advantageous instrument.
Especially in such companies with a lack of venture capital, decisions to
invest in new, often expensive manufacturing structures must be
safeguarded by methodical and reliably accurate planning aids.

The experience gathered in a number of case studies has been used
repeatedly in a series of similar problems presented to the Fraunhofer
Institute for Production Engineering and Automation (IPA), Stuttgart,
for evaluation by different, mostly medium-sized, companies. The IPA
has been intensively involved since the mid-1970s in the design and
development of 'MUSIK', a modular simulation system for flexibly
linked manufacturing systems, which has now been successfully used to
help in the design of flexible manufacturing systems in more than a
dozen planning tasks.

Simulation as an indispensable tool in detailed FMS planning

The increase in planning uncertainty and the investment risk thus
entailed which develops parallel to the growth in the flexibility of
processing, material flow and information flow in automated manufac-

turing systems can only be eliminated by regarding components such as production cells, transport facilities and storage devices as parts of an integrated overall system from the outset rather than as isolated elements.

Utilisation of simulation results for planning and control tasks

The variety of planning and design problems to be expected in view of the large number of different types of systems under consideration meant that the idea of general applicability had to be allowed for in the development of 'MUSIK'. The modular structure devised with this aim in mind enables the widest range of production, material flow and control structures to be simulated and studied on the computer model, irrespective of the transport principle, variety of orders and number of planned machines in the system. By combining different software modules and parameter constellations, 'MUSIK' permits the definition of the main points of investigation adapted specially to the planning phase and particular considerations involved. The mutual dependence of different, often parallel running, production and material flow processes and their dynamic nature can thus be registered in detail and evaluated, not only for planning purposes but also for later disposition and control tasks (Figs. 1, 2).

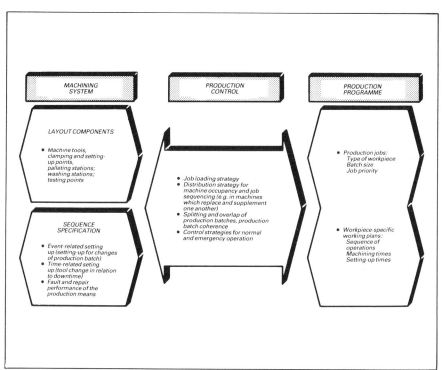

Fig. 1   'MUSIK': specific possibilities for representation, model components and data structures (part I)

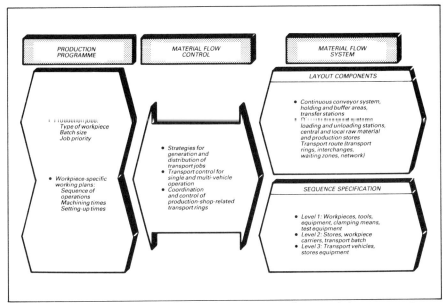

Fig. 2 'MUSIK': specific possibilities for representation, model components and data structures (part II)

The integrated use of 'MUSIK' in several consecutive planning phases shown in Fig. 3 permits the planning results to be gradually refined and adapted so as to provide reliable quantitative decisions at the variant choice and rough conception stages. On the basis of the model designed for this purpose, additional modules and design forms

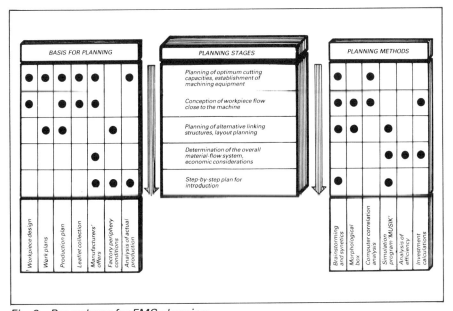

Fig. 3   Procedures for FMS planning

oan bu aotivated to show specific parameters hitherto not considered and thus to plan and optimise special production processes in detail. The possibility of successively integrating specific problem-related disposition and control aspects into the model also provides a well-founded basis on which to make decisions at the implementation and installation phases of such complex manufacturing systems.

Degree of detail in model formation as decisive factor

In every simulation the quality of the results obtained depends on the degree of detail attained at the model formation stage. Fig. 4 shows the complexities in design and development involved in order to simulate the characteristics of all or nearly all of the system constellations in full detail in the model. This diagram only sketches a small selection of the different material handling concepts and their effect on the possible supply and disposal strategies.

The different hardware concepts involved in material handling necessitate the precise definition of special strategy modules to be combined in varying degrees with one another during the model formulation phase so as to be able to simulate the decentralised control logic of the transfer process at every station in the required detail. The desired emphasis in the master transport control configuration to be devised for the FMS simulator in accordance with the specific application can be obtained from the overall logic derived from the strategy components.

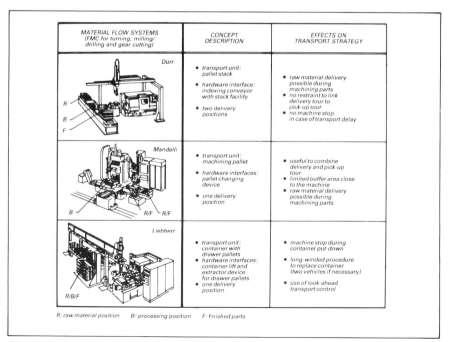

Fig. 4   *Different concepts of material handling*

Case study 1: FMS for prismatic engine parts

The flexibility of an already extensively designed manufacturing system for Adam Opel AG, Rüsselsheim, for machining cylinder heads, was to be examined during the planning stage. The drop in demand of almost 50% for cylinder heads in the engine model on which the concept had originally been based had made it necessary to extend the workpiece spectrum. To enable cylinder blocks to be machined on the same line alongside the different types of cylinder head it was also essential to revise the entire production line concept.

FMS studies

Fig. 5 shows the final layout of the most recently implemented manufacturing system[1]:

- In the planned production system two types of cylinder head were to be completely manufactured in a first (outer machine ring) and second (inner machine ring) clamping, and, in addition, final machining of cylinder blocks was to be incorporated.
- The machining system considered of a total of 14 machining centres, a precision milling unit and a continuous wash station. The machining of the different types of workpieces was to be carried out in a non-specific (chaotic) sequence. A special coding and identifying system automatically diverted the workpiece to the relevant station and also called up the desired machining program.
- The workpieces entered the production system via three separate clamping and set-up stations and ran through, clamped on workpiece-specific pallets, in an anticlockwise direction. For machining purposes, the individual workpiece carriers were diverted to the right or left from the central transport system (endless, continuously revolving special double-chain belt) and delivered via feed devices to the appropriate target station.

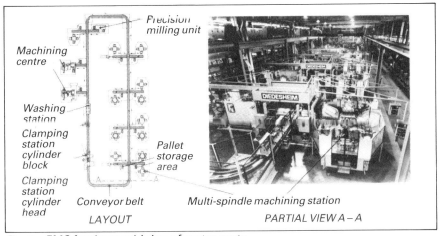

Fig. 5   FMS for the machining of motor parts

The machine tool company commissioned to plan and implement the system was aware that for reasons of the complexity involved the detailed planning of this manufacturing concept could not be carried out and adapted correctly without simulation. The main emphasis during this planning phase was a simulation-assisted sensitivity analysis of reductions in productivity by taking account of planned and random machine downtimes.

Determination of the influence of different tool-changing strategies on the system

At a fixed production interval, in this case every 300 cylinder heads, the individual machines had to be closed down to differing degrees so as to replace worn tools. With details concerning the tool life (300, 600, 900, 1,200, 1,800, 2,400 and 3,600 workpieces) of all the tools used in the line, the organisation of the tool-changing system for all 350 tool spindles distributed on approximately 40 cassettes could be worked out. Fig. 6 shows how long, in relation to the finished workpieces, the individual machines had to be shut down for retooling purposes and hence how many spindles and cassettes were affected. To keep the loss of productivity resulting from the tool-changing procedures down to a minimum, an optimal sphere of responsibility for each of the planned staff of workers, in this case two or four, had to be defined and special retooling sequences determined.

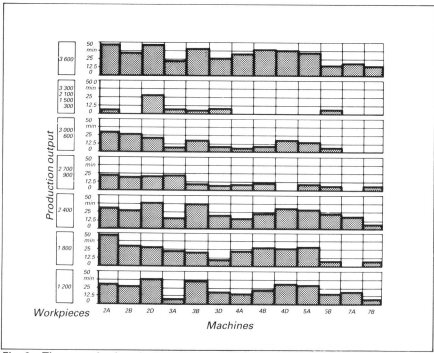

Fig. 6   Time required to change tools depending on production input

The flexible continuity afforded by the transport system and the additional waiting stations designed as workpiece buffers meant that the individual workstations were relatively independent so that planned machine downtimes for retooling purposes could be bridged in some cases.

Improved retooling organisation and personnel utilisation

The simlation results shown in Fig. 7 clearly reflect the increased flexibility. The individual retooling sequences, during which the workers replaced the worn tools successively on the machines in their charge, meant that at no time were all workstations simultaneously stopped. After a complete retooling cycle of 3600 parts (i.e. all tools in the system have been replaced at least once), the loss in productivity in relation to the planned tooling personnel capacities was 5% for two persons and 1.5% for four persons.

It is not until the manufacturer has figures such as these that he can make a well-founded decision as to the correct tool-changing strategy, bearing in mind the costs for tooling personnel and the loss in utilisation through the relative reduction in productivity.

Development of emergency strategies as a reaction to machine breakdowns

Even in a manufacturing system as flexible as the one described here, with no redundancy in the machining system at all, local disruptions can lead to a total breakdown of the organisationally and physically linked system. It is nevertheless very important both for the running-in stage and later in the continuous operation of newly installed manufacturing systems to prepare strategies to ensure first that those machines not affected by a breakdown continue to be supplied with workpieces for as

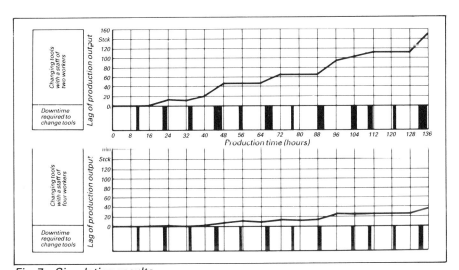

Fig. 7   Simulation results

long as possible, and secondly to provide the best possible conditions for re-establishing full operation in the shortest time after the machine breakdown has been rectified.

The following emergency strategies were illustrated in the model, and with the aid of simulation their relative effectiveness was shown and compared:

- Emergency strategy 1 – Recursive buffering of up to three workpiece pallets per work unit, dependent on length of disruption.
- Emergency strategy 2 – Free passage of workpiece pallets past station affected by breakdown.

Emergency strategy 1, in which a maximum of three pallets per station are backed up recursively from the failed station, led very quickly to a blockage of the entire system. Circulating workpieces which had already been machined at the failed station could be further processed and leave the system. All other workpieces running chaotically through the system were backed up between the clamping station and the failed station. In the case of a breakdown at station 5, for example, the entire engine block production came to a standstill although only station 8 was still required to complete machining.

The flexibility provided by the transport system and individual waiting stations is much better utilised by emergency strategy 2. The individual workpiece pallets can be conveyed onwards, irrespective of the duration and location of the breakdown, and be further processed enough for the missing operations to be completed as soon as the failed station is back in commission. Fig. 8 shows these results expressed as a capacity utilisation curve of all the machines integrated in the system before, during and after the rectification of the relevant machine

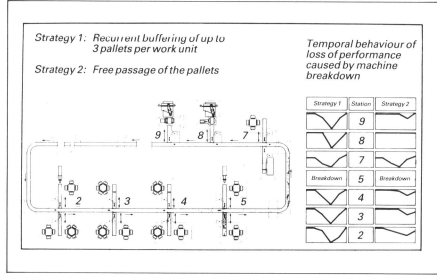

Fig. 8 *Simulation-based investigation of different emergency strategies*

breakdown. It can be seen that the utilisation of the stations not affected by the breakdown drops much more slowly with strategy 2 than with strategy 1. All machines except those in station 7 could continue to be supplied during the breakdown with restricted performance so that they could at least be partially utilised.

Case study 2: FMS for rotating parts from the production range of a car component supplier

To ensure the long-term competitiveness of machined parts production in a supply company to the motor vehicle industry and to eliminate the technical and organisational weak points the entire rotating parts production sector was to be comprehensively restructured as part of a projected investment programme.

Planned FMS

The flexible manufacturing system devised by the IPA in conjunction with the user and designed to assure the time- and cost-saving production of rotating parts provided for the use of autonomous, easily retooled CNC machine tools which, in view of the extreme fluctuations in the demand structure and the planned successive introduction of an additional type of unfinished part by the end of the 1980s, were to be linked into a flexible material flow system.

Fig. 9   Container lift and extractor device for drawer pallets

The implementation of three-axis loading gantries or palletting robots, depending on the accessibility and the existing external measuring facilities, was planned for the individual production machines for the purposes of automatic workpiece handling. Detailed simulation studies showed that, because of the sometimes very short machining times, an adequate supply of workpieces to the individual stations could only be assured with the use of discrete conveyors by transporting several pallets together, and the individual transportation of flat pallets could only be considered if continuous conveyors were used[2]. Detailed material flow examinations showed that the rigid linking of the individual machine tools via roller tracks, for example, would not permit the high flexibility needed as a result of the varying machining processes. With account taken of the technical and financial criteria involved, a mobile transport container with retractable drawer-type pallets was proposed for conveying the workpiece batches. In order to be able to standardise the different interfaces the Euro-pallet format (1,200 × 800mm) was chosen, with horizontal-axis workpieces being positioned in prism-shaped mounting strips with length adjustment. It was proposed that the transport containers be automatically transferred by discrete transporters, such as inductively guided vehicles, to the lifting and extracting device, which would have to be installed at every machine so that the individual workpieces could be brought to the handling area of the loading gantry, for example (Fig. 9).

Requirements and problems of simulation studies

In all simulation studies the reproducibility of the results must be assured so as to be able to make a comprehensive comparison of the economic and organisational criteria of the planned system and existing production. A realistic model of future production processes cannot be obtained, however, unless the organisational requirements and peripheral conditions used in the simulation are the same as those to be used in the computer-assisted master control system. Despite the capabilities of the standard simulation modules implemented to date, the model in this case had to be adapted to allow for the following special features:

- Parallel machining of workpieces in a production batch on a maximum of two machines in one operating sequence.
- Simultaneous machining of workpieces from a maximum of four different production batches on the machines in one operating sequence.
- Maintenance of production batch content during entire production run.

Dimensioning and design of material flow system

The requirement for a continuous and uniform transport and supply concept for all rotating parts production operations threw up two rival

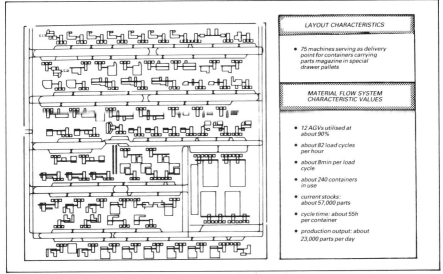

Fig. 10   *FMS for rotational parts: proposed system I*

and sufficiently flexible material flow variants:

● Material flow variant I (Fig. 10) – Transport and direct transfer of workpiece containers to lifting and extracting mechanism at each machine by inductively guided vehicles alone.

● Material flow variant II (Fig. 11) – Transport of workpiece containers between different manufacturing lines and transfer to factory-related linear buffers by inductively guided vehicles. Preparation and transfer of individual drawer-type pallets to pallet cycling devices at each machine by rail-bound transport carriers.

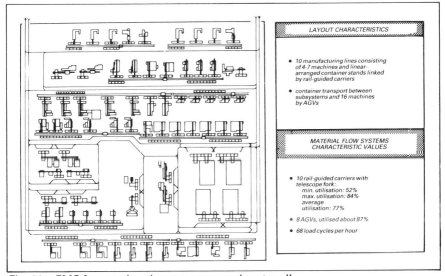

Fig. 11   *FMS for rotational parts: proposed system II*

Material flow variant II with several autonomous production lines, comprising individual machines, rail-bound transport carriers and a linear buffer for containers opposite the machine line, was a much less convincing solution than variant I, despite the apparently attractive flow structure. With the aid of the simulation it was shown with great clarity that the effect of reducing the investment cost through the overall material handling system in a common linear buffer was overestimated at this stage in the planning. The order structure, batch size distribution and special cycle time characteristics prevented the inclusion of even larger manufacturing lines which would have been necessary to ensure the economic feasibility of this material flow variant. Fig. 12 shows characteristic simulation values which had to be determined individually for every separate part system in the overall layout before an investigation of the material flow system as a whole was possible.

If these figures are examined in greater detail it can be seen that they are not sufficient on their own by any means to build up another manufacturing line not yet simulated. Manufacturing line III, in which almost all parts are friction welded on a total of seven machines, had to be built up in such a way that approximately 7,900 workpieces per day could be handled. Despite this relatively small amount the corresponding transport carriers were utilised almost as much as in production line II, which was responsible for machining almost 13,000 workpieces per day on five machines in two clamping operations.

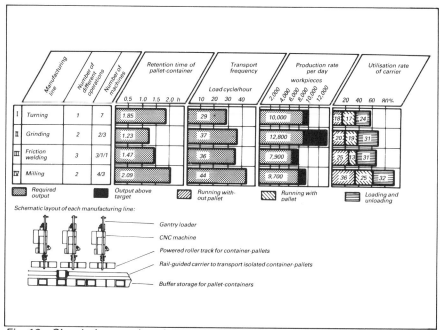

Fig. 12   *Simulation results: characteristic values of different manufacturing lines with pallet supply by means of rail-guided transport carrier*

Case study 3: FMS for rotating parts from the production range of a manufacturer of precision parts

The technical feasibility of a completely revised work sequence for machining precision parts in the framework of a comprehensive restructuring of the corresponding factory could not examined to the full, with the result that for a certain time at least two manufacturing sub-areas had to be considered with differing degrees of automation of the workpiece handling and preparation operations:

- Sub-area I comprised operations such as turning, drilling and milling for which proven CNC machine tools with relevant automation periphery were readily available on the market. Three-shift operation, as required for economic reasons, was already feasible, given the current state of the art. In addition, operations such as internal and face grinding were to be integrated. From the point of view of automation, considerable improvements were still required, particularly with regard to workpiece handling, but they were by no means insurmountable. The machining centres with pallet changers selected for the milling/drilling work were integrated for reasons of autonomy in an independent small flexible manufacturing system with material supply by means of a rail-bound transport carriers.
- Sub-area II was and will be characterised by operations such as broaching, honing and deburring for which no increase in the degree of machine automation at an acceptable cost can be anticipated in the foreseeable future and for which existing conventionally operated one- or two-shift operation was planned.

Integrated consideration of conventional and highly automated sub-areas

Detailed simulation-assisted material flow studies showed that pallets would be the most suitable medium for the internal transport and machine positioning of workpieces not only in the automated but also in the conventionally operated production area. Because of the special batch size characteristics with less than 20 workpieces per batch, almost all production orders could be magazined on a half-size Euro-pallet (800×600mm). As part of the gradual introduction of the new system it was planned that the transport of parts to the different areas of the factory should also be automated, and here, too, there was a clear indication of the advantages from the point of view of flexibility of inductively guided vehicles.

In regarding the system as a whole it was essential to illustrate both manufacturing areas in the same model. As they would be controlled in future by a common computer system this presentation was necessary in order to make decisions as to investment costs. Fig. 13 shows a highly simplified version of the simulation layout used for demonstrating the structure of the machining and material flow system.

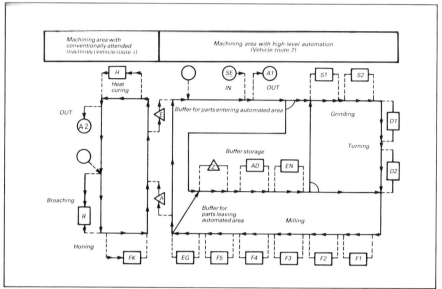

Fig. 13   Two machining areas with a different degree of automation: simulation layout

By considering the overall system as an integrated unit on the model it was possible to draw conclusions at any time about the special characteristics of each individual area, particularly the automated one (Fig. 14). In retrospect it appeared highly practical for inductively guided vehicles to take over transportation in the conventional area.

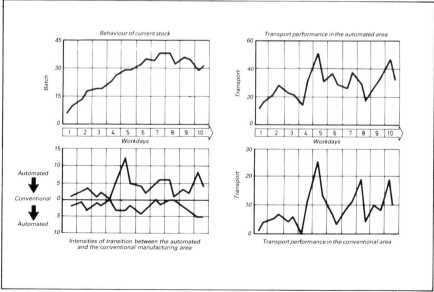

Fig. 14   Material flow parameters for a manufacturing system with differently automated subdomains

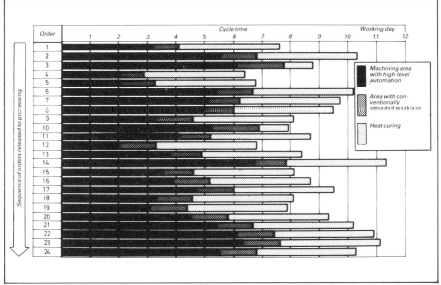

Fig. 15   Dynamical behaviour of different portions of cycle time

Despite the relatively long batch machining time and the small amount of transportation involved it was still possible to achieve almost 60% utilisation. In view of this utilisation value the inclusion of an unfinished parts store and heat treatment in the material flow system was also considered.

The improvements to be expected in the delivery capabilities and flexibility as a result of the restructuring of the production sector as a whole was impressively documented in the distribution of the measured cycle time (Fig. 15). If it can be assumed that the cycle time for heat treatment will remain constant in the long term, the future rationalisation potential for the overall machining system can be considered as exhausted, at least as far as automation is concerned.

Allowance for machine breakdowns

A widely used method for determining technical and time capacity requirements to be allowed for machine downtimes is to define a useful running factor of, say, 0.8 for machines in two-shift operation. This means implicitly that 20% of the theoretically possible machine utilisation time is reserved for breakdowns and cannot be used for machining. Because of its static character, however, this correction factor does not always correspond accurately enough to conditions in reality. Machine breakdowns tend to be haphazard and do usually cause unforeseen long interruptions.

A breakdown profile for intensive drilling/milling machine stations evaluated specifically for this planning task from empirical observations also had to permit the study of the behaviour of the independent small FMS during dynamically changing interruptions (mean time to repair,

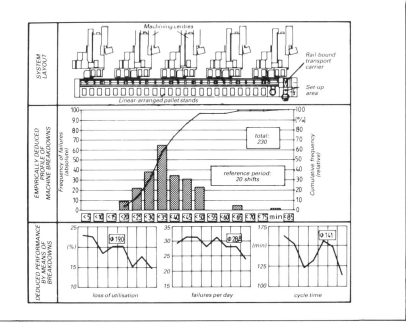

Fig. 16   Availability analysis with simulation support

MTTR) and interruption-free intervals (mean time between failures, MTBF). To this end, distribution functions were derived the statistical parameters of which reflected the course of the breakdown profile.

A simulation-assisted sensitivity analysis of these parameters within the limits provided by the breakdown profile produced, as expected, no significant reduction in the output of this part manufacturing system over an extended observation period. In Fig. 16, however, it is very easy to see that the production process was subject to haphazard fluctuations which would have to be detected by an intelligent production control system and compensated through appropriate disposition and control measures.

Concluding remarks

Because of the number of possible variants involved, the individual design and adaptation of conceivable alternative concepts required for objective system comparison (differing technologies, machine concepts, work procedures and linking sturctures) can only be reliably achieved within an acceptable period of time with the aid of computer simulation. This does not produce an optimal analytical solution but rather a satisfactory solution achieved through iteration. Profitability figures, such as material rates, productivity, cycle times and system availability, which could scarcely be calculated without this tool and could thus only be roughly estimated, make it possible to select a tailor-made quantitatively assured overall system for future automated production. The consideration of different machining sequences, product variants and batch sizes simultaneously provides the basis for well-founded decisions as to the optimal degree of flexibility. This means that at an early stage in the planning different conceivable peripheral considerations and conditions can be included with the aid of the computer.

References

[1] Wolf, G. (Ed.) 1984. *Modulares, flexibles Maschinenkonzept für Kleinund Mittelserienfertigung*. Kernforschungszentrum Karlsruhe GmbH, KfK-PFT-E 16, February 1984.
[2] Warnecke, H.-J. and Steinhilper, R. 1983. Neue Entwicklungen zur flexiblen Automatisierung der Teilefertigung. *VDI-Z*, 125(20): 853-859.
[3] Mills, R.I. and Talvage, J.J. 1985. Simulation programs for FMS design. In, *Proc. 1st Int. Conf. on Simulation in Manufacturing*, Stratford-upon-Avon, UK, March 1985.

DECISION SUPPORT FOR PLANNING FLEXIBLE MANUFACTURING SYSTEMS

G. Seliger, B. Viehweger and B. Wieneke
*Institut für Produktionsanlagen und Konstruktionstechnik (IPK),
West Germany*

With the increasing complexity of modern manufacturing and assembly processes the number of alternatives to be analysed during the planning process increases. To help planners to analyse solution alternatives quickly, different computer-aided planning tools have been developed. The building-block-orientated modelling system MOSYS incorporates a mathematical-analytical method and simulation. The mathematical-analytical method is used for roughly estimating the performance of a host of different FMS concepts. It is based on the theory of queuing networks, with knots representing elementary order queues and their connections standing for order paths. Simulation enables the dynamic behaviour of those FMS concepts selected by the mathematical-analytical method to be analysed in detail. For the specific task of designing flexible manufacturing systems with integrated tool handling, the parametrised simulation system TOSYS has been developed. The structure of these planning tools is described and illustrated by examples of industrial applications.

Complex manufacturing and assembly processes require advanced modelling techniques for system structure analysis in design and implementation phases, and for scheduling and control once production systems have been realised. In principle, computer aided design and planning aids will enable the generation, valuation and performance assessment of more solution variants in a shorter planning time, thus boosting human creative engineering. Decision support for planning tasks must meet the requirement of stepwise refinement in system design. Fig. 1 shows application areas for different planning tools.

Mathematical-analytical methods are suitable for the quick analysis of many solution alternatives in the phase of rough planning. The

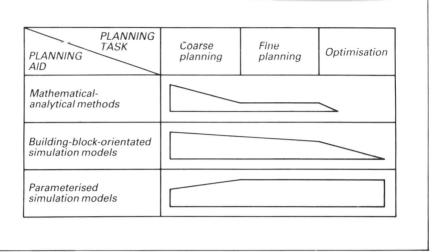

Fig. 1   Application areas of planning aids

values and sensitivity of the parameters of the manufacturing system can be determined by a few experiments. Thereby certain alternatives can be excluded from further planning in this early stage. Other alternatives become more detailed in the next phases of the planning process and are evaluated by means of simulation experiments. As regard simulation building-block-orientated simulation systems are more flexible, whereas parameterised simulation systems enable more detailed modelling.

Planning tool MOSYS

Structure

For the complex tasks in planning manufacturing systems the planning tool MOSYS has been developed. This interactive decision support system is based on a description system that has been derived from system-theoretical considerations and incorporates two analysis methods: the mathematical-analytical method and simulation (Fig. 2).

The description is made up with five building blocks:

- Manufacture.
- Assemble.
- Transport.
- Test.
- Store.

These building blocks are adjusted to the specific requirements by using their particular attributes (Fig. 3). Modelling of the system structure and input of the routings for the parts, pallets and tools is supported by graphics. Further data, such as the distance matrix, set-up matrix, production programme, input order and system initialisation, are given to the planning system via a user-friendly dialogue.

Fig. 2 Planning system MOSYS

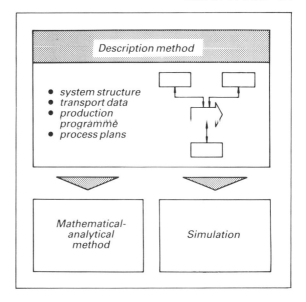

In addition, the control of the manufacturing system has to be described. This can be done either by explicit programming of the rules or by a synchronisation mechanism based on a semaphore concept. As the building blocks have standardised interfaces for the control, they can be combined freely to include any of these alternatives, thus enabling the modelling of a variety of control strategies by simple means.

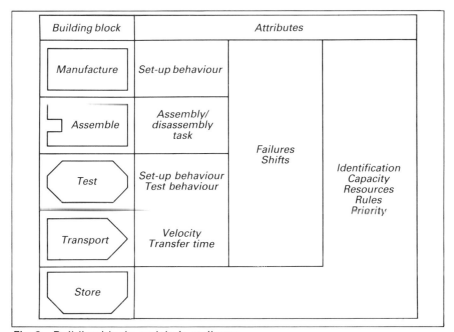

Fig. 3 Building blocks and their attributes

The planning system MOSYS has been written in SIMULA and is presently running on DEC 2060 hardware.

Mathematical-analytical method

The mathematical-analytical method (MAM), incorporated in MOSYS, is based on the theory of queuing networks. Queuing networks are graphs where knots represent elementary order queues and their connections stand for order paths. Queuing networks are simple models to determine the characteristic values of a system.

Fig. 4 shows the description of a manufacturing system as a queuing network. The modelling of any processing unit and any transport system is possible. The modelling is done by the description part of MOSYS. The description data are prepared and can be modified during a dialogue. The following characteristic values of the manufacturing system are calculated:

- Characteristic filling f^*.
- Bottleneck of the system.
- Maximal throughput of the system.
- Actual filling.

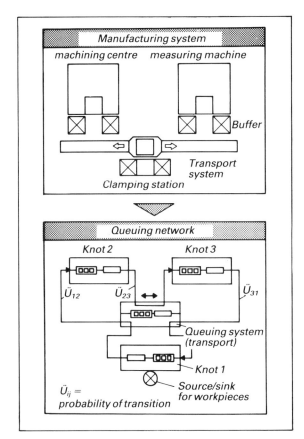

Fig. 4 Modelling of a manufacturing system as queuing network

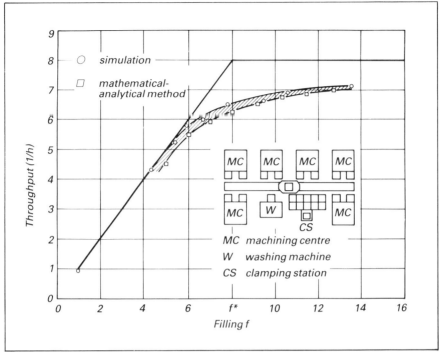

Fig. 5 Characteristic curves

The actual filling of a system is the number of orders which are in the system on average. If the system filling exceeds the characteristic filling f^* this leads to longer waiting queues in front of the processing units, and therefore to longer turn-round times, without increasing the system throughput significantly. If the workpieces are on pallets the filling equals the number of workpiece carriers used in parallel.

For each single building block the following values are calculated:

- Maximal throughput.
- Utilisation.
- Queuing length in front of the building block.

Additional outputs are the characteristic curves of the analysed manufacturing system (Fig. 5). These curves give some information about the behaviour of the system when changing the production requirements.

Experiments have proved that the utilisations of the capacity units calculated by MAM differ maximally by 10% and on average less than 5% from simulation results. If utilisation of the capacity units is less than 90%, the difference for maximal system throughput, filling, turn-round time and queue length is less than 10% (Fig. 6).

The descriptions of the solution alternatives that are used for MAM can be detailed in the further planning process and can be used for simulation experiments. This decreases modelling effort.

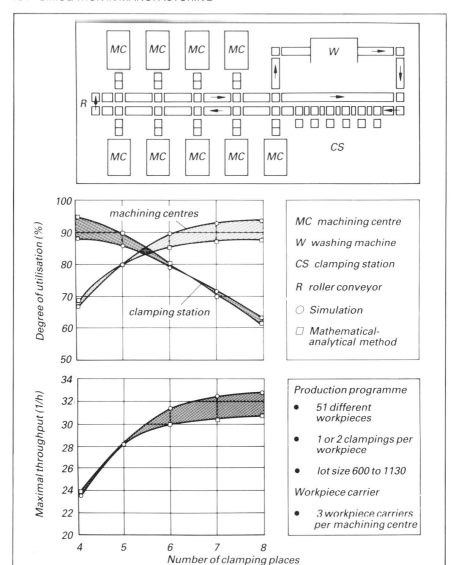

Fig. 6   Mathematical-analytical method for FMS with continuous conveyor system

Application examples

In the following, two applications of the MOSYS system are described. The first one is a large assembly system which has been modelled on a rather rough level in order to determine the better layout alternative and optimal input scheduling. In the second case a flexible manufacturing system had to be modelled in detail to get more exact results regarding system behaviour in the case of machine failures. In both cases the hierarchic modelling capabilities of the MOSYS system have been utilised.

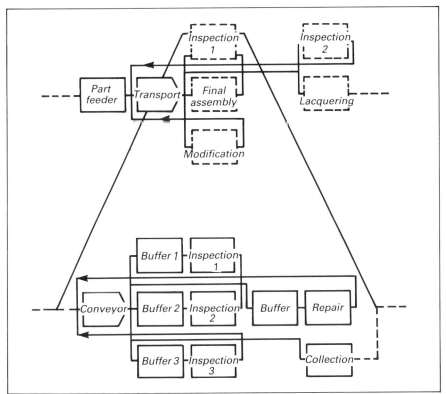

Fig. 7   Hierarchical modelling of flexible assembly system

Assembly line for printed circuit boards. In many companies of the electronics industry, rationalisation efforts concentrate on the assembly areas for pcbs. In these assembly areas electronic components are manually or automatically placed on the pcbs, soldered, tested and repaired. In this case study two different layouts had to be analysed. The alternatives differed in the number of supplementary and substituting workplaces, set up requirements and interlinking transport systems. The goal of the simulation analyses was the determination of the better alternative and system behaviour in relation to different order input strategies.

The system is subdivided into two sections with assigned transport systems. The first section includes about 150 manual workplaces and automata; the second section comprises about 170 testing stations, repair places and other manual workplaces. The second section as modelled by the description part of MOSYS is shown in Fig. 7. The customer gave typical production programmes, set-up behaviour and failures of the capacity units. The pcbs are transported in magazines, each containing up to 12 boards. 2500 pcbs have to be produced per day with about 40 different order types. The necessary number of capacity units was estimated by MAM and had to be modified only in a few cases due to dynamic influences.

Simulation experiments for the chosen alternative showed that a steady input in small lot-sizes in mixed order types gives a balanced system behaviour. Other results of the simulation were:

- Turn-round times for each order type.
- Utilisation of automata and manual workplaces.
- Set-up times for each capacity unit.
- Behaviour in the case of machine failures.
- Buffer dimensioning.

Flexible manufacturing system. For a planned FMS for motor parts, simulation experiments should help to improve system behaviour. This large FMS consists of several flexible manufacturing cells for boring and milling processes. These cells are interlinked by an automated guided vehicle (AGV) transport system. Each single cell has its own transport systems for workpiece handling. The first step in the simulation was the analysis of the single cells. The following example of such a single manufacturing cell, shown in Fig. 8, consists of:

(MC)	Five CNC machining centres in groups of two and three machines.
(WM)	A washing machine.
(CS)	A clamping station with transport device.
(WC)	A transport system.
(SP)	Workpiece storage places.

The parts are machined in two clampings. Each pallet carries one workpiece. About 100 jobs have to be produced in two shifts.

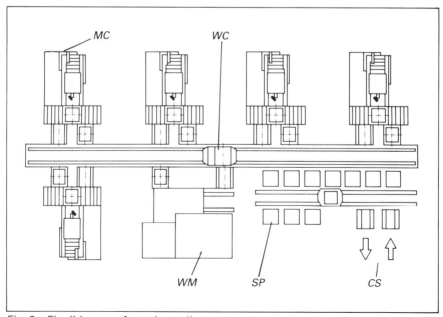

Fig. 8   Flexible manufacturing cell

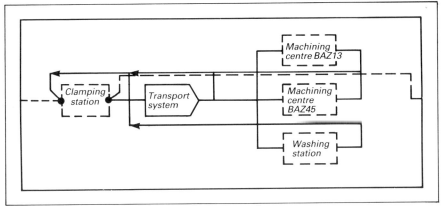

Fig. 9   Modelling of flexible manufacturing cell

Machine down-times were modelled to obtain more detailed results regarding:

- Throughput.
- Turn-round time and waiting time for the orders.
- Number of pallets needed.
- Number of buffers and storage places.

The basic data for the machine down-times were the results of an analysis of failures of NC machining centres over a period of four months. From these data the MTBF (mean-time-between-failures) and MTTR (mean-time-to-repair) have been derived.

The model of the flexible manufacturing cell (Fig. 9) was built up from the subcomponents:

- Clamping place with transport system, buffers for pallets and unclamped workpieces (Fig. 10).
- Both machine groups with buffers.
- Cell transport system.

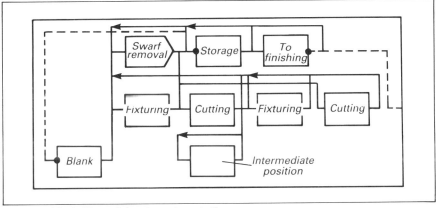

Fig. 10   Modelling of the clamping station

The inclusion of an extra transport device at the clamping station helped to avoid potential bottlenecks in the transport system. The grouping of the machining centres provides a better utilisation of the system in the case of machine failures.

Simulation system TOSYS
Structure

The parameterised simulation system TOSYS has been developed for the analysis of flexible manufacturing systems with automated tool flow and the associated strategies. Manufacturing systems with separate workpiece and tool transport systems can be analysed. The exchange of tools between the tool magazine and the spindle and the transfer of tools to and from the central tool storage, which are executed in parallel to each other, are modelled.

The adjustment to different system configurations is done by parameter variation. The workpiece transport system can be AGVs or tracked vehicles. Besides machining centres, measuring machines, deburring stations, washing machines, clamping stations and storage places can be modelled. The tool transport system can be robots in overhead design or tracked vehicles. Fig. 11 shows the structure of the simulation system. The parameters of this system are:

- Number and arrangement of processing stations and storage places.
- Number of places in the central tool magazine.
- Acceleration and velocity of the workpiece and tool transport systems.
- Assignment of transport vehicles to certain stations.
- Transfer time for pallets.
- Grip time for tools.

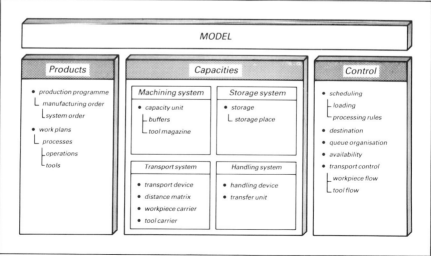

Fig. 11   Structure of the simulation system TOSYS

- Number of places on a tool vehicle.
- Number of places in the tool magazine.
- Turning time of the tool magazine.
- Tool exchange time.
- Strategies for tool exchange.
- Work time regulations for operators and machines.

The results of simulation runs are given as tables or Gantt charts via monitor or printer. Simulation results are:

- Utilisation of stations.
- Empty run, load run and transfer times of the transport vehicles.
- Number of pallets needed.
- Turn-round time for orders, subdivided into its components.
- Work time of the operator for clamping of workpieces and input/output of tools.

With parameterised simulation systems like TOSYS a high degree of exactness of the description, and therefore a high accuracy of the simulation results, is achieved. This is connected with the disadvantage of less flexibility in its application.

Application example

For a flexible manufacturing system with automated tool flow, different variants had to be analysed by simulation experiments using TOSYS. The system (Fig. 12) consists of:

(MC)	Eight CNC machining centres.
(WM)	A CNC washing machine.
(CS)	Four clamping and set-up stations.
(TIO)	A tool set-up station.
(WC)	A tracked workpiece transport system.
(TC)	A tracked tool transport system.
(TS)	Four tool storages, each with 90 places.
(–)	60 workpiece storage places (not marked in the figure).

The two alternatives, which are explained in the following, differ in the number of places in the tool magazine.

In alternative A the tool magazines of the machining centres have 40 places. With the given production program this causes the exchange of five tools on average when the order type changes at the machine. These exchanges have to be done by the tool transport system in addition to the exchange of worn-out tools. Simulation analyses had to show whether the tool transport system is a bottleneck and how big the availability reduction of the machining centres caused by this bottleneck might be.

For alternative B the tool magazines were enlarged to 60 places so that all tools for the production programme are available at the machines. The tool transport system is needed only for the exchange of

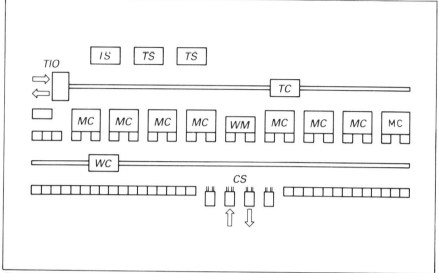

Fig. 12   Flexible manufacturing system with automated tool flow

worn-out tools. Simulation had to clarify whether the workpiece transport system was becoming a bottleneck.

Fig. 13 shows that the tool transport system in alternative A is totally overloaded because of the frequent order changes at the machining

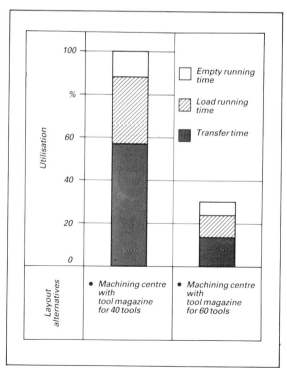

Fig. 13   Utilisation of tool transport system

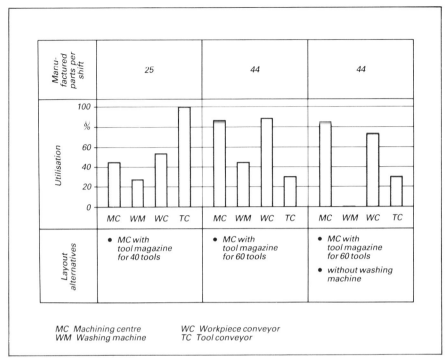

Fig. 14   Utilisation of the system components

centre. As there are only a few empty runs no effect can be gained by empty run minimising strategies. The utilisation of the machining centres is low because tools are often not in the magazine when they are needed (Fig. 14). Although the workpiece transport system in alternative B has a very high utilisation of more than 90% there is nearly no reduction in utilisation for the machining centres. This is because of the relatively long processing times per workpiece, so that it is nearly always possible to supply the machining centres with workpieces in time.

Concluding remarks

The costs of an analysis of planned systems are determined by expenditure for:

- Data preparation.
- Modelling.
- Simulation runs.
- Interpretation and documentation of the results.

Both systems discussed in this article have been able to decrease the necessary efforts for modelling, which is often the most time-consuming part of the analysis. The building-block-orientated planning system MOSYS provides more flexibility in modelling, but has the disadvantage of less accuracy in special cases. Parameterised simulation systems like the TOSYS system for FMS with automated tool flow support a high degree of detail for a small spectrum of applications.

References

[1] Viehweger, B. and Wieneke, B. 1986. Rechnerunterstützte Planungshilfen für Fertigungssysteme. *ZwF*, 81(1): 23-28.

[2] Viehweger, B. 1986. *Planung von Fertigungssystemen mit automatisiertem Werkzeugfluß*. Reihe Produktionstechnik Berlin. Carl Hanser Verlag, München.

[3] Spur, G., Viehweger, B. and Wieneke, B. 1983. Problem oriented methods for planning and optimization of flexible manufacturing systems. In, *Proc. 2nd Int. Conf. on FMS*. IFS (Publications) Ltd, Bedford, UK.

[4] Seliger, G. and Wieneke, B. 1984. Analytical approach for function oriented production system design. *Robotics & CIM*, 1(314): 307-313.

[5] Seliger, G. 1983. *Wirtschaftliche Planung automatisierter Fertigungssysteme*. Reihe Produktionstechnik Berlin. Carl Hanser Verlag, München.

[6] Engelke, H. et al. 1985. Integrated manufacturing modelling system. *IBM Journal of Research and Development*, 29(4).

'FORSSIGHT' AND ITS APPLICATION TO AN FMS SIMULATION STUDY

M.J. Birch, T.J. Terrel and R.J. Simpson
Lancashire Polytechnic, UK
and
H.P. Feszczur
British Aerospace plc, UK

A model of a flexible manufacturing system may be developed by applying the simulation package FORSSIGHT to a previously derived structured requirement. A chosen sub-system of the FMS is represented by a dynamic mimic diagram on a colour graphics display to provide interactive simulation facilities for the user. The modular approach used throughout all stages of the model development is described, and it is shown that this form of flexible simulation is able to accommodate all phases of plant automation.

British Aerospace is presently engaged in the evaluation of large-scale flexible manufacturing systems within its production facility at Strand Road, Preston. Because of the complexity of these systems, a simulation study is essential before the project team can make confident design decisions.

When commencing a major simulation study it is important to carefully assess the methods and tools that are available. First, the use of a definitive requirements methodology is desirable as it provides a visible, unambiguous simulation requirement for the user. The methodology should encourage the formation of a modular structure within the analysis, thus providing flexibility for future modifications and developments. CORE has been adopted as the requirements methodology[1].

Secondly, a simulation package is required which is based on a language in common use, can be block-structured to aid coding from the CORE requirement, and has effective monitoring facilities for rapid debugging, as simulations tend to be more difficult to validate than conventional programe. The package should provide good facilities for the processing and output of statistical data over long simulation runs. The capability of displaying sub-systems dynamically as interactive mimic diagrams is a useful additional facility.

The CORE methodology

CORE is a semi-automated, structured analysis and design method-ology which hierarchically decomposes the system into processing modules and then checks the validity of each module. Comprehensive documentation is produced with a common notation employing process/data interaction diagrams in a time-ordered sequence.

The simulation package FORSSIGHT

Many computer simulation languages exist (e.g. GPSS, ECSL, SEE-WHY, HOCUS, MAST, AUTOMOD, SIMSCRIPT, SIMON, SIMULA, FORSSIGHT). FORRSIGHT[2] possesses all the required properties and has been adopted for the simulation of the Treatments Cell, although it is recognised that many of the other packages may also be suitable.

The package may be implemented on any computer that has standard FORTRAN IV and COBOL compilers. It was developed by the Operational Research Dept of the British Steel Corporation[2]. FORSSIGHT comprises two distinct areas:

- *FORtran Simulation System (FORSS)*: This is the basis of the simulation package. It is non-interactive and non-visual and provides a comprehensive statistical analysis of monitored data in tabular and histogram form on completion of the simulation run.
- *FORSSIGHT*: This is a powerful further development of FORSS which is used to drive interactive colour mimic diagrams for the simulation model.

FORSS

FORSS is a discrete-event package which assumes that any system is constructed of uniquely identifiable units known as 'machines' which proceed through discrete 'states' as system time advances. When a machine changes from one state to another an 'event' occurs; time is the only changing factor between events. If a unit does not experience events it is not a machine. In the case of automated FMS applications, 'machines' represent programmable logic controllers which in turn drive the physical devices.

Prior to entering the first cycle of a simulation run, the system passes through an initialisation stage during which run data and starting conditions can be assembled according to the user's instructions.

Descriptions of events are coded as B(ound) or C(onditional) event blocks. A B-event is an event bound to take place at a given time with a given machine, whereas a C-event takes place only on condition that certain specified circumstances hold (e.g. machine available, item present in queue). FORSS passes through three phases in each time cycle:

- The A-phase advances simulated time to the next event, which occurs when a busy machine becomes available.
- The B-phase scans through the machines to identify those that become available at that time. For each such machine FORSS generates an entry to the appropriate B-event block or, in the case of a null event, identifies the machine as 'available' at that time.
- The C-phase considers each C-event in turn, checking whether the conditions governing that event can be executed. When all the C-event blocks have been scanned, and where possible executed, all the necessary changes within the simulation at that time have been completed. The A-phase is then re-entered for the next time advance.

An additional task of the A-phase is to check if simulated time exceeds desired run-time. When this condition is satisfied, the system enters a termination stage in which user-defined results may be calculated and output.

FORSSIGHT

The basic FORSS package is enhanced by provision for the on-line display of colour mimic diagrams during a simulation run and for on-line interaction with the model by the user.

The facilities include the simulation of continuous movement of items on a display and the control of the relationship of the display time-scale to real-time. Multiple displays may be manipulated in a single simulation run, permitting the study of transitory states of a system, different parts of a system or the same part at different levels of detail.

The general FORSSIGHT screen layout includes facts panel, interactive window, time panel and display (Fig. 1). The facts panel contains information relating to the specific run, the interactive window presents input/output data, the time panel indicates simulated time and the display area contains the dynamic, visual representation of the model.

FORSSIGHT has extensive interactive facilities. For example, at any time during a run the current screen may be stored or parameters altered. Time is initially set within the model but can be adjusted from the keyboard by the user during a run. The user can suspend or end operation, or direct an entry to a specific model block to exchange data interactively. Statistics collected in distribution streams can be updated dynamically in histogram form within the display area. On completion of the run, the terminal is left with the final display retained.

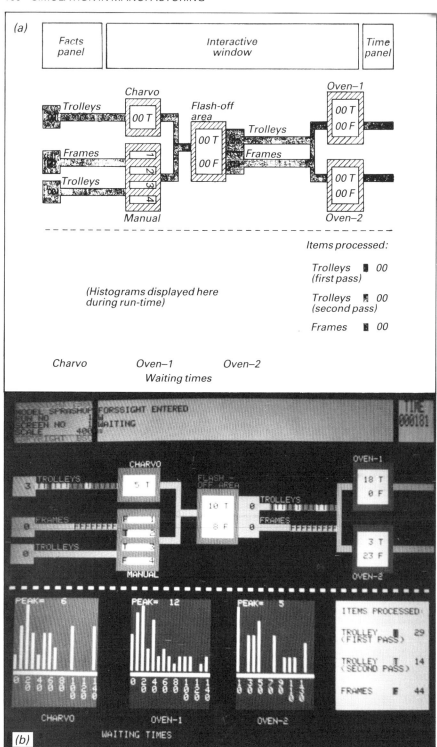

Fig. 1   (a) Screen layout for paint shop mimic diagram, and (b) terminal display

Application of FORSS to the simulation requirement

A CORE requirements analysis was carried out on the Treatments Cell for Aluminium. A queue-driven modular construction results in which individual machines or machine-groups operate as self-contained sub-systems. Any given sub-system can enter into an active state provided that it is 'opened' by a shift-scheduler, the machine is not busy and an 'item' is available on its queue of work. Once a machine completes a process or process-chain it is once more made available for work. It is evident from reference to the previous section that this description of the requirements analysis also bears a close relationship to the FORSS event block structure.

During the phased implementation of the Cell, process additions and modifications are required which must first be reflected in structural changes to the CORE requirements. These changes are then implemented within the FORSS model by adding event blocks or modifying existing blocks. The model is then re-run and the output assessed. By designing a loosely coupled system with well-defined interfaces, maintenance costs are minimised since modules may easily be modified in isolation.

The FORSS source code for the robot sprayer (or 'Charvo') is shown in Fig. 2. The full-scale model employs a more complex scheduling system but the basic philosophy remains the same: machines or machine-groups operate as individual system modules, decoupled by their input and output queues.

Suppose the C-phase has been entered and block C107 is being processed. If the paint shop is in operation, if the Charvo is available (i.e. not loading, spraying or unloading) and if a trolley is present in the queue, then the Charvo commences loading the trolley. The load time is determined from a negative exponential distribution (as are all process times for this model) and the Charvo is committed to loading the trolley for this period.

The input queue volume is then monitored, the trolley removed from the queue and its waiting time in that queue monitored. The trolley is added to the Charvo, which is incremented as a result. The Charvo volume and capacity are checked for equality. If the Charvo is full, it commences spraying (B108) on completion of the loading period. Otherwise it is made available for further loading via the null block (B107).

When the Charvo enters the spraying block its volume is sampled. Spraying continues for 480 time units. It enters the unload block on completion of spraying.

When the Charvo enters the unload block the time required to unload the Charvo is determined from the product of the time to unload a trolley and the number of trolleys in the Charvo. The Charvo is then committed to unload its contents for this period, the trolleys being moved individually into the flash-off queue. The Charvo capacity for the next load is set up and the present Charvo volume set to zero.

```
C107                                        LOAD-TROLLEY
C
      INTEGER CHLDTM,MSPRCH
C DEPART FROM  EVENT BLOCK IF CRANE OR TROLLEY NOT AVAILABLE
      IF ((.NOT.A(107)).OR.(CHTLVL.EQ.0)) RETURN
C TIME TO LOAD TROLLEY SAMPLED FROM NORMALISED DISTRIBUTION
      CHLDTM = 0
      WHILE (CHLDTM.LE.0) DO
            CHLDTM = NEG(117,120)
      ENDWHILE
C COMMIT CHARVO FOR DURATION OF LOADING
      CALL D(CHLDTM,107)
C SAMPLE THE QUEUE VOLUME
      CALL DIS(CHTLVL,113)
C TAKE THE NEXT ORDER FROM THE QUEUE
      CALL DROP(CHTRLS)
C DECREMENT THE INPUT QUEUE VOLUME
      CHTLVL = CHTLVL - 1
C MEASURE THE WAITING TIME AND PLACE ON THE SAMPLE STREAM
      CALL MVCAR(ATTRIB,QS,MSPRCH,0,1)
      MSPRCH = (TIM - MSPRCH)/60
      CALL DIS(MSPRCH,112)
C PLACE THE ORDER ON THE CHARVO AND INCREMENT THE LOAD COUNT
      CALL PUSH(QS,CHARVO)
      CHARVL = CHARVL + 1
C COMPARE CHARVO VOLUME AND CAPACITY
      IF (CHARVL.LT.CHARCP) THEN DO
C IF CHARVO NOT FULL, RE-ENABLE FOR NEXT LOADING
            CALL E(107,107)
      ELSE DO
C IF CHARVO FULL, BEGIN SPRAYING
            CALL E(108,107)
      ENDIF
      RETURN
      END
B107                              RE-ENABLES CHARVO
      CALL D(0,107)
      CALL E(0,107)
      RETURN
      END
B108                              CHARVO-SPRAYING
      INTEGER CSPRTM
C PLACE CHARVO VOLUME ON SAMPLE STREAM
      CALL DIS(CHARVL,119)
      CSPRTM = 480
C COMMIT CHARVO FOR DURATION OF SPRAYING
      CALL D(CSPRTM,107)
C ON COMPLETION, BEGIN UNLOADING (B109)
      CALL E(109,107)
      RETURN
      END
B109                              UNLOAD-CHARVO

      INTEGER CHULTM,SSPRCH
C TROLLEY UNLOAD TIME SAMPLED FROM NEG EXP DISTRIBUTION
      CHULTM = 0
      WHILE (CHULTM.LE.0) DO
            CHULTM = NEG(120,120)
      ENDWHILE
C TOTAL TIME TO UNLOAD DERIVED FROM CHARVO VOLUME
      CHULTM = CHULTM * CHARVL
C COMMIT FOR DURATION OF UNLOAD
      CALL D(CHULTM,107)
C MOVE TROLLEYS FROM CHARVO TO "FLASH-OFF" AREA UNTIL CHARVO EMPTY
      WHILE (CHARVL.GT.0) DO
            CALL DROP(CHARVO)
C RE-SET THE QUEUE ARRIVAL TIME
            SSPRCH = TIM
            CALL MVCAR(SSPRCH,0,ATTRIB,QS,1)
            CALL PUSH(QS,SPTRLS)
            CHARVL = CHARVL - 1
      ENDWHILE
C MAKE CHARVO AVAILABLE FOR WORK
      CALL E(0,107)
C RE-SET THE CHARVO CAPACITY FOR NEXT LOADING
      CHARCP = 0
      WHILE (CHARCP.LE.0) DO
            CHARCP = NOR(118,4,1)
      ENDWHILE
C SET CRANE EMPTY FOR START OF NEXT LOADING
      CHARVL = 0
      RETURN
      END
```

Fig. 2   FORSS source code for 'Charvo' module

Finally, the Charvo becomes available for work when unloading is complete.

Though the intra-block processing is less complex than the complete simulation, it serves to emphasise the modular philosophy discussed at the beginning of this section. The addition of further machines (e.g. transport functions) is simply achieved by including further modules as required.

Graphical simulation of the paint area using FORSSIGHT

The existing Treatments area has been modelled in FORSS, providing extensive statistical data on queue volumes and waiting times (see Fig. 3). The automation of the Treatments facility is now in progress, the initial area of interest being the paint shop. This presently comprises manual sprayers, the Charvo, a flash-off area and two ovens. In this version of the model, transportation is a manual function and is assumed to be permanently available.

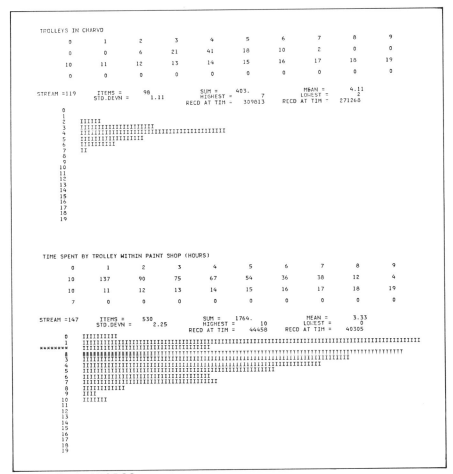

Fig. 3   Typical FORSS output

It is important that engineers should be able to rapidly assess the impact of changes in the equipment specification and layout of the paint shop. A visual and interactive simulation package, such as FORS-SIGHT, is effective in enhancing the already extensive FORSS results by providing a powerful, graphical 'what-if' facility. It is also useful for demonstrating proposed systems to both management and users.

The screen is set up during initialisation by using the static routines. FORSSIGHT permits a maximum of 20 dynamic symbols in a single display. Each symbol is referenced by an identifier. Thus the location of frames and trolleys to be moved need not be known. Facilities also exist for changing the position and/or size of otherwise static display areas, such as the linear queues. The screen layout is shown diagrammatically in Fig. 1a and is illustrated in Fig. 1b.

Trolleys are queued at the Charvo, and trolleys and frames at the manual spray area. Buffers are provided which increment if the linear queues overflow. The total number of trolleys loaded in the Charvo is displayed, with the indicator 'SPRAYING' if B114 has been entered. Individual frames and trolleys are shown within the four manual spraying areas.

On completion of spraying, trolleys and frames move instantaneously to the flash-off area, where they wait for 15 minutes before proceeding to the oven queues. The dynamic mechanism of the oven and spray queues are identical. Oven–1 and oven–2 are loaded with trolleys and frames according to an algorithm compiled from shop-data. The number of trolleys and frames loaded are shown within each oven area.

Future trends

The Treatments simulation represents the state of this current methodology to date. The C/B block structure models both the physical activities of the machines and, by default, the software logic which controls these activities. One possible way forward is to separate the two areas (physical activities and control logic) and in so doing produce generalised, possibly parameter-driven modules for both. This will enable a specific simulation to be constructed for any FMS requirement from a library of standardised modules based on the control software. The methodology will also provide a CORE control software requirement, the resultant code and the simulation environment within which to test this code. Hence, by utilising the control logic as a basis, a single requirement can be produced for both simulation and control software. The results of initial work confirm the feasibility of this approach.

Concluding remarks

With the continuing increase in the complexity and cost of flexible manufacturing systems, it has become essential to adopt simulation techniques both to aid design decisions and to reduce the difficulties encountered during user familiarity. However, problems exist concerning time-scale and manpower requirements. The use of a standardised routine library can assist in reducing these overheads.

The current modular simulation analysis techniques discusssed here, combining both CORE and FORSSIGHT, are proving to be an extremely useful tool for the development, demonstration and maintenance of automated manufacturing systems at British Aerospace (Warton Division).

Acknowledgements

The authors are grateful to the Manufacturing Systems Department of British Aerospace (Warton Division) for giving its full support to this project and to British Aerospace plc for permission to publish this paper.

References

[1] Birch, M.J., Terrell, T.J. and Simpson, R.J. 1984. CORE Methodology and its Application to the Simulation of an Automated Manufacturing Facility. September 1984.
[2] British Steel Corporation (Operational Research Dept) 1981. Report TM20/2 (Issue 4), FORSSIGHT User Manual, September 1981.

INTRODUCING FMS BY SIMULATION

A.S. Carrie and E. Adhami
University of Strathclyde, UK

A project simulating a flexible manufacturing system being installed in a collaborating company is presented. The objectives include verifying the supplier's performance predictions, assessing the sensitivity of system performance to variations in data, and development of operating policies. Several models permitting various aspects to be examined were written using the ECSL language. The activity cycle diagrams are described, and assumptions and results discussed. A significant finding was that the system could become blocked, and suggestions were made to the supplier for software revisions to handle this problem.

A collaborating company is installing a flexible manufacturing system (FMS) for the manufacture of complex castings, with a gross cost of over £7 million. With an investment of this magnitude, a company wants to be certain that target levels of output will be achieved. At the tendering stage, firm operational data is in short supply, for the simple reason that until detailed part programming has been done operation times are only estimates. Consequently management also requires to know how sensitive to variations in part mix, operation times, machining methods and so on will be the operating performance of the system. Simulation is an appropriate method of evaluating these parameters, and in this case the supplier presented simulation results to support the system design and performance predictions. At a very early stage in the project the authors were asked to undertake simulation studies of the proposed system, to assess these factors and to verify the predictions given by the supplier.

Product data

The castings weigh up to 2.5 tonnes, and require between 80 and 209 tools in their machining. With conventional NC methods the lead time was around three months. The parts concerned were of two types, booms and gearboxes. Some 13 parts were specified as candidates for the FMS in view of their complexity, size and future requirements. The operations involve the full range of work carried out on machining centres, and a special feature of several of the parts was the need for back facing operations, in which the cutting tool has to be inserted manually into the holder after the holder on the spindle has been positioned through a bore beyond the face to be machined.

Process planning

It was recognised that the operations requiring manual intervention would not be suitable for carrying out on a machining centre under CNC control unless at the expense of poor utilisation and productivity. An early decision was that there should be a special machine in the system where all such operations would be carried out. This meant that the remaining operations could be done at normal machining centres. Decisions were necessary as to the number of these required and how the operations should be allocated. In view of the large number of tools needed for each part and of the total number of parts, these were complex and interrelated decisions.

Of the 13 parts designated for the FMS, seven were to be loaded on the system at start-up and the other six introduced over the following 12 months. As part of its design study the supplier carried out a detailed investigation of three of the first seven parts and the results were extrapolated to the remaining parts.

The operations involved were classified into three groups: roughing, semi-finishing and finishing operations. Because of the complexity of the parts it was decided that up to three fixtures would be required for each part. In some cases the part would need resetting within the same fixture.

System design

The system, shown in Fig. 1, comprises five live spindle CNC horizontal boring machining centres and one horizontal boring machining centre with a facing head for the operations requiring manual intervention. All the machines will have a tool changer with a tool magazine of capacity 100 tools. Two machines will be tooled similarly for roughing operations. The other three will be tooled for semi-finishing operations, but due to the number of tools involved not all operations may be done at more than one of these machines. Castings will be fixtured and moved on pallets by an automated guided vehicle (AGV) following a wire buried in the floor. At the load/unload area and at each machine there will be two pallet stands acting as buffers between the AGV and the machine table. There will be 13 pallets. The system will be controlled by a DEC PDP 11/44 executive computer, which will route the parts through the system, schedule the machining operations, monitor tool life and store part programs. It will also produce management reports on the status of the system.

Further details of various aspects of the system have been given elsewhere[1].

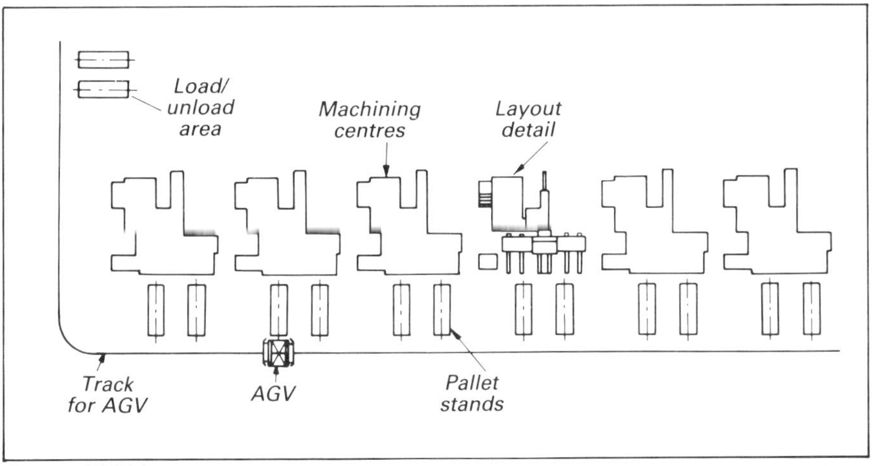

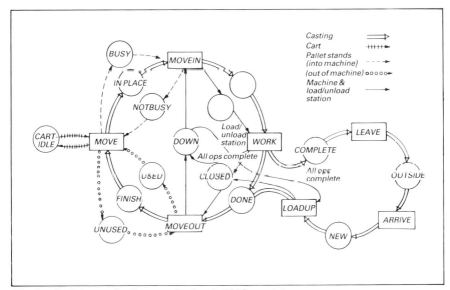

Fig. 1   FMS layout

Initial system model

At an early stage in such a project many details have yet to be resolved. The model-builders were faced with the fact that very few details were then available about how the parts would be fixtured and palletised. Nor was it known whether there was any restriction on the use of the pallet stands in front of each machine. By making some judicious assumptions it was possible to arrive at a very simple model of the system. Fig. 2 shows the activity cycle diagram of the model. Such diagrams, or their equivalent, are used in several simulation languages to show what the various entities in the model do and how they interact.

Fig. 2   Activity cycle diagram for basic FMS model

In this case the language in which the model was written was ECSL (Extended Control and Simulation Language)[2].

The principal assumptions are:

- One pallet stand will be used for parts queuing for a machine, and the other will be used by parts waiting to be moved from the machine after machining. This simplifies the logic of queues and pallet stand availability, and is in fact the same as in the supplier's model and in the executive computer software, although there is no physical restriction on the use of the stands for either purpose.
- The load/unload area can be regarded in the same way as any other work station, except that new raw castings arrive here from outside the system, and that completed ones leave it at this point. This permits the loading, unloading, resetting within the same fixture and fixturing operations to be defined as part of the operation sequence and handled like any other operation.
- The availability of fixtures, pallets and tools will not influence the system performance (there being no precise details available). These entities can therefore be omitted from the model.
- The system does not suffer breakdowns. Therefore any entities and logic to handle breakdowns can be omitted from the model.
- Because of the batch nature of the operations there will be only one set of part types in the system at any time, and a batch of raw castings of each of these types will be available at the load/unload area at the start of a batch run. This means scheduling rules for new work can be omitted from the model.

In Fig. 2 the queue OUTSIDE represents the rest of the company's environment. Activity ARRIVE is the delivery of a batch of booms and gearboxes, which wait for loading onto pallets in queue NEW. Queue DONE represents a casting which has been processed by a machine, or loaded onto a pallet at the load/unload area, waiting to be moved from the machine table onto the outgoing pallet stand, in the off-queue. This is done in activity MOVEOUT, which transfers the casting into queue FINISH, where it awaits the AGV moving it to the next required work station. Activity MOVE is the move, and following it the casting is put into queue INPLACE (i.e. on an on-queue pallet stand). Activity MOVEIN is the casting being moved onto the machine table which should be immediately followed by activity WORK, the processing of the casting by the work station. If a casting is at the load/unload station at the end of its operation sequence, then instead of entering the queue DONE again it is placed in COMPLETE to LEAVE the system. The cycle for the machine tool (or load/unload station) is from the queue DOWN (i.e. waiting for work), via activities MOVEIN and WORK into queue CLOSED, where it remains until the casting is moved out from the machine table onto the off-queue pallet stand. In the case of the load/unload station, the station can also go from DOWN to LOADUP when a new casting is available, and for this reason it is convenient to separate LOADUP from the other work at the load/unload station. The cart is IDLE when not engaged in activity MOVE. The on-queue pallet stand is converted from NOTBUSY to BUSY by activity MOVE which brings a casting to it, and back to NOTBUSY by activity MOVEIN which puts the casting onto the machine table. The off-queue pallet stand is made UNUSED by MOVE and USED by MOVEOUT. This representation of MOVE implies that the off-queue and on-queue stands are engaged for the entire period of the move and not just at the beginning and end moments respectively. This simplifies the logic and will not be unreasonable if the move times are short, but could be altered if necessary.

The data used by the model consist mainly of the sequence of operations, including load, reset, refixture and unload, with times for each. Other data state how the machines are grouped (e.g. which are the roughing machines), and give estimates of the time to move pallets between pallet stands and a machine and between work stations.

No operation times were available for the facing head machine, so this machine was omitted, as it had been in simulation runs done by the supplier using the MAST model[3].

Results

One immediate effect was to find that the system could become congested, with it being impossible to move castings away from the load/unload area because the on-queue pallet stands at the required machines were already occupied by castings. These castings could not go onto the machine because the machine table was occupied by a casting waiting to go to the off-queue pallet stand, which was itself occupied by a casting awaiting moving to the load/unload area. To resolve this problem activities TAKEOFF and RELOAD were added to the model whereby, when this situation arose, the casting at the load/unload station would be taken off, and laid aside in a queue LAYBY until the congestion is relieved. It was also found that the

Table 1   Results from supplier's, basic and improved models

	Supplier's model	Basic model	Improved model		
			number of castings in system:		
			6	9	13
Production rate (parts/week)	28.2	28.2	28.4	28.4	28.4
Mean time in system (min)	1144	2006	935	1151	1276
Mean % utilisation of:					
Roughing machines	77	78	78	78	78
Finishing machines	60	64	64	64	64
Load/unload station	28	37	37	41	47
AGV	13	64	64	65	65
Number of 'take-offs'	–	808	6	68	185

Based on 8 weeks' operation, omitting facing head

frequency of this event could be reduced by changing the cart's priorities. It was suspected that this would be highly dependent on the number of castings allowed into the system. The improved model therefore included a limit on the number of castings in the system. It also included fixtures, on the assumption that only one fixture of each type would be available. Table 1 shows a summary of these results.

These results show the effect of the number of castings in the system on the time in the system, the utilisation of the load/unload station and the number of take-offs. They also show good correlation with the supplier's results, with the exceptions of the load/unload and cart utilisations, which are affected by the simplifying assumptions.

Revised model

This question of congestion of the system and the need to take castings out of the system focused attention on the need to define the operating procedures more clearly, particularly as the company did not wish to lift pallets out of the system to avoid the risk of damage to locating edges. It was thought that the assumptions in modelling the load/unload area were contributing to the problem, and that in a more detailed model the situation might not arise. At this time more data became available about the travelling speed of the AGV, distances involved, procedures at the load/unload area, and so on. It was therefore decided to construct a more comprehensive model.

Fig. 3 shows the part of the activity cycle diagram for the load/unload area, and Fig. 4 the part of the diagram for the machine area. Fig. 3 shows substantial revision, while Fig. 4 shows only minor variation from Fig. 2 .

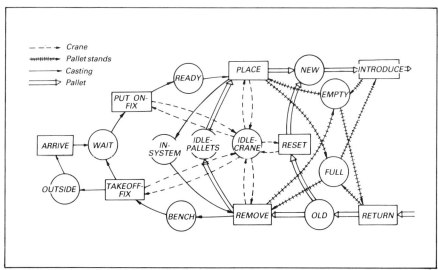

Fig. 3   Activity cycle diagram for load/unload area

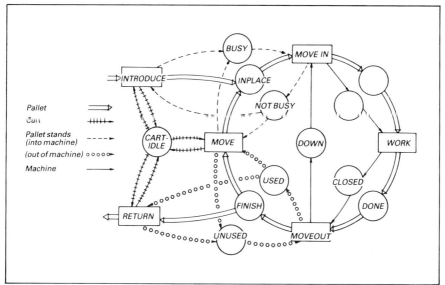

Fig. 4   Activity cycle diagram for machine area

Castings arrive (activity ARRIVE) from outside the system (queue OUTSIDE) as raw castings and wait for processing (queue WAIT). First the casting is put in a fixture with the aid of an overhead crane (activity PUTONFIX) and is ready (queue READY) to be placed (activity PLACE) on a pallet at one of the two pallet stands at the load/unload area. Once placed on a pallet the casting is regarded as being in the system (queue INSYSTEM) until brought back to the load/unload area. Then it is removed from the pallet (activity REMOVE), placed on a work bench where it waits (queue BENCH) to be taken out of the fixture (activity TAKEOFFIX). After being taken off the fixture the casting may be completely machined, in which case it is taken away to the assembly area (queue OUTSIDE), or else it may require further machining on a new fixture, in which case it rejoins queue WAIT. In the ARRIVE activity logic compares the number of parts of each type completed with the number scheduled and generates a casting of the type with the lowest ratio, thereby ensuring balanced part production.

Fixture availability must be checked before the casting can be put on the fixture. A variable defines how many fixtures of each type are available for use. When a casting is put on its fixture the number of fixtures of that type is reduced by one, and when it is taken out of the fixture the number is increased by one. By this means, formal representation of the activity cycle of the fixture does not need to be included in the model.

Once a fixtured casting is placed upon it, a pallet waits (queue NEW) to be moved to a machine by the AGV (activity INTRODUCE), where it is placed on the in-queue pallet stand (queue INPLACE) at a suitable machine. The pallet then follows the cycle for the casting in Fig. 2, through MOVEIN, WORK, DONE, MOVEOUT to queue FINISH, where it waits for the AGV to move it elsewhere. The move may be direct to another machine (activity MOVE), or it may be back to the load/unload area (activity RETURN). A pallet brought back waits on a pallet stand (queue OLD) for service. Either the casting and its fixture are lifted off the pallet (activity REMOVE) or the casting is reset within the same fixture (activity RESET).

Movement times were computed by using the distances involved and the speed of the AGV, these figures now being available. As in the initial model, all moves by the AGV require that the destination pallet stand must be vacant. At the machines one stand is an on-queue and one is an off-queue. At the load-unload area each pallet stand fulfils both roles. Activities MOVE and INTRODUCE take a NOTBUSY on-queue pallet stand and put it into queue BUSY, while MOVEIN takes it from BUSY and puts it into NOTBUSY. Activity MOVEOUT requires an UNUSED off-queue stand and converts it to USED, and activities MOVE and RETURN return it to UNUSED. At the load/unload area, activities PLACE and RETURN take a stand from EMPTY into FULL, with INTRODUCE and REMOVE doing vice versa.

The cycle for the machines is similar to that in Fig. 2, with the omission of the involvement in LOADUP.

The operators at the load/unload area have not been included in the model since all the activities there involve the overhead crane, and since there is only one of those this is the limiting factor. The crane is idle when not involved in PUTONFIX, PLACE, REMOVE or TAKEOFFFIX.

Model extensions

The model was extended in various ways to permit particular questions to be examined. These extensions have been omitted from Figs. 3 and 4 for the sake of clarity.

Empty pallets

Fig. 3 shows that there is a stock of idle pallets (queue IDLEPAL-LETS) which are placed onto a pallet stand or removed from it. This implies that empty pallets are kept off the system at the load/unload area. However, as mentioned above, the company do not intend to remove empty pallets, but rather to move them to vacant stands within the system. The model was therefore a simplification, but provided a yardstick for comparison with the supplier's model, which made a similar assumption. As a matter of fact, the executive software also assumes that empty pallets will be out of the system. This matter has been referred back to the supplier.

The inclusion of empty pallets in the system necessitates new logic for moving them around from one stand to another, frequently to vacate a stand required for a loaded pallet to be brought to a machine. The logic can become quite complex. The method adopted is to observe the stand which is the destination of a move a few minutes before the move should be made, and if it is occupied by an empty pallet to move the pallet out of the way. If this is done too far in advance other moves may intervene, and if too late the move may be delayed. Several time values were evaluated to find the one which minimised the number of moves.

Blockages

The problem of congestion observed in the initial models was found to recur, and was not limited to the load/unload area. The situation could arise where there were three loaded pallets at two machines, one on each of the machine tables, and none could move because the destination pallet stand was occupied. The pallet at the off-queue of each machine needed to go to the on-queue of the other machine. Logic had to be added to the model to unblock the system whenever this occurred. The question of how the real system would cope with blockages was referred to the supplier.

It was decided that logic should be incorporated to prevent blockages occurring. It was suggested that if there were never more than two loaded pallets at any station this would be achieved, and logic to this effect was added. Blockages occurred even quicker.

Non-dedicated pallet stands

A study of the situation when blockage occurred suggested that the problem might be avoided if pallet stands could serve as either on-queue or off-queue, not just one or the other. Logic to permit this was incorporated and found to be effective, although there was a marginal reduction in output of the system.

The logic was also used in evaluating the effect of pallet stand breakdowns.

Vacuum swarf removal

Swarf is a practical problem since chips can remain inside the casting and interfere with subsequent operations. It was proposed to provide a robot vacuum cleaning station, where a vacuum tube controlled by a robot could be inserted inside the casting to remove chips. Three possible locations were evaluated. Two of these, at the load/unload area and at a separate station with the casting on the AGV, required no further logic, since the effect could be achieved by adding to the load/unload or transport times. The third possibility, to incorporate an extra pallet stand where the cleaning would be done, required extra logic somewhat similar to that applied to the stands at the load/unload area.

Table 2   Comparison of initial and revised models

	Initial model	Revised model
Production rate (parts/week)	30.2	32.2
Mean % utilisation of:		
Roughing machines	87	89
Finishing machines	66	71
Load/unload station	44	19
AGV	69	21

Based on 4-week runs with 13 castings permitted in system

Results

A comparison of the initial model and the revised model, including blockage-relieving logic but excluding empty pallets and the facing head machine, is given in Table 2. This shows slightly improved output with the revised model. It highlights the effect on the load/unload station of eliminating take-offs. It shows that the utilisation of the AGV has been substantially altered by the more accurate transportation time algorithm.

When approximate time estimates for the facing head machine operations became available, the data were added to the model. Table 3 summarises the results of several runs. The results show a general reduction in output and a substantial increase in time to put castings through the system, relative to the initial model.

Empty pallets

Run 1 is from the model in which empty pallets were removed from the system, while run 2 is from the model including them. This shows that the AGV is some 25% more utilised, while output and machine utilisation are reduced by about 10%.

Table 3   Results of simulation

	Run number									
	1	2	3	4	5	6	7	8	9	10
Production rate (parts/week)	18.5	17.1	14.5	10.5	16.8	16.5	20.1	18.3	17.9	17.2
Mean time in system (min)	2631	2715	2852	4418	2780	2787	2216	2605	2694	2781
Mean % utilisation of:										
Roughing machines	84	80	96	52	78	76	91	84	82	80
Finishing machines	65	61	52	38	60	59	72	65	63	61
Facing head machine	83	78	65	97	76	75	90	83	81	78
AGV	35	44	37	29	42	43	44	35	34	33
Crane	24	22	19	14	22	23	25	24	23	23
Pallets	55	53	55	54	55	52	81	58	56	54
Fixtures	36	37	50	31	43	38	33	39	37	35

Based on 10-week runs

Variations in operation times

When production methods were planned in greater detail it was decided that an additional roughing operation might be needed. Run 3 shows that the effect of a 75% increase in roughing times would be a drop in output of 15% due to the very high utilisation of the roughing machines.

At this time the facing head operation times were only very rough estimates, and company engineers feared these might double when tapes were proved out. Run 4 shows that the effect of a 100% increase in facing head times would be a substantial drop in output of 60% relative to run 2, with the facing head machine becoming a serious bottleneck. To relieve this it might be necessary to work this machine in three shifts, with accompanying organisational problems.

Vacuum swarf removal

In run 2 vacuum cleaning is done on the AGV. In run 5 it is on the extra pallet stand, and in run 6 it is done at the load/unload station. The system is relatively insensitive to the changes.

Fixture availability

Fixtures are expensive and only one of each was proposed. However, in order to see whether this would restrict the system, run 7 was done assuming two sets of each fixture. System output improved by some 18%, suggesting that additional fixtures would be advantageous. The utilisation of some fixtures was very low and therefore only some should be duplicated.

Reliability

The supplier's model assumed that the system would be operational for 75% of the possible hours, and down for 25%, and for comparison this assumption was also made. The company had several CNC machines from the supplier and their reliability was good, so this was considered a reasonable assumption. However, the reliability of the pallet stands was unknown, so runs were carried out to assess this factor. Runs 8, 9 and 10 assumed failure of one pallet stand at a roughing, finishing or facing head machine respectively from the third week of the run to the end. These runs included the logic to permit stands to be either on-queues or off-queues, and because of this additional complexity excluded empty pallets. The new logic adds flexibility counteracting the loss due to the stand failure. The reductions compared are of only a few percent – up to 7% in the case of failure of a stand at the facing head machine.

Accurate operation times

As planning proceeded, operation times were revised and the models rerun from time to time to assess the effects. These runs showed a steady deterioration in output due to lengthening operation times. In August 1981 the average work content per casting was estimated at 624 minutes, and the production capacity of the system estimated at approximately 1140 parts per annum. By September 1982 the work content had risen to an average of 809 minutes per part, and the system output had dropped to 920 parts per annum. By June 1983, when NC tapes had been proved out for the first three parts, the work content became 907 minutes. When all parts have proven tapes the models will be rerun. A further drop in output is inevitable, and this is a matter for some concern.

Concluding remarks

The simulation project has provided a variety of useful results so far, and is expected to do so in the future as the system is commissioned. The main overall conclusions are:

- Although the models were written in a different language from the supplier's, with a different way of describing the real world, and with different sets of constraints and implicit assumptions on the way a model is constructed, there was broad agreement between them.
- The work has found two problems which could seriously affect the effectiveness of the real system, and the supplier has been asked to look at them. These are blockages and empty pallets. Suggestions were made to the supplier based on the logic found effective in the model. There was at first some difficulty in convincing the supplier that the system could become blocked, and his response to these is awaited.
- Initial estimates of operation times and system capacity, including those in the system proposal which led to the contract, have been rather optimistic.

Attention is now being switched to operation aids. It is planned to develop a short-term scheduling aid which will take the tool preparation requirements into account. Since the first seven parts require almost all of the tool magazine capacity this will be an important aspect when new parts are added to the system. It is hoped that this software will be linked with the FMS executive computer and the factory production computer, and will provide colour graphical output.

References

[1] Carrie, A.S., Adhami, E., Stephens, A. and Murdoch, I.C. 1983. Introducing a Flexible Manufacturing System. In, *7th Int. Conf. on Production Research*, University of Windsor, Canada, August 1983.
[2] Clementson, A.J. 1982. *ECSL User's Manual*. CLE-COM Ltd, Birmingham, UK.
[3] Lenz, J.E. 1980. *MAST User's Manual*. CMS Research Inc., Oshkosh, Wisconsin, USA.

EXPERIENCE IN THE USE OF COMPUTER SIMULATION FOR FMS PLANNING

T.C. Goodhead and T.M. Mahoney
University of Warwick and Austin Rover Group, UK

It is widely appreciated that simulation fulfils the role of quantifying facility requirements and evaluating alternative operating strategies to ensure that the required system performance may be achieved in an economic manner. However, a simulation model can also make highly significant contributions in the following areas. It imposes a discipline which tends to improve awareness of all factors having an effect on system performance. It aids the understanding of the complex inter-relationships which arise as a proposed method of operation is developed. It acts as a focal point and a communications channel for all involved personnel and may form the basis for contractual agreement between customer and supplier. To illustrate these points, the use made of a computer simulation package with graphics and interactive facilities during the planning stages of large-scale flexible manufacturing systems is described.

Most of the problems currently experienced with existing flexible manufacturing systems (FMS) derive from a lack of detailed pre-planning which results in a poor system specification being written[1]. This leads to problems arising between potential FMS users and equipment suppliers at the contractual agreement and installation stages, with the outcome that a system often does not perform as was intended, it costs more than expected, and it is late.

The specification for an FMS may be poor for a number of reasons. It may be:

- Inaccurate, in so far as it defines a system which simply will not operate in the intended manner.
- Sub-optimal, in that it is grossly inefficient in terms of utilisation of resources and is therefore more expensive than necessary.
- Incomplete, in that not all details which can in any way affect the

performance of the whole system are considered, even though the details which are included are accurate.
- Misunderstood, so that even if it defines a satisfactory system and includes all details which affect the system, it is open to misinterpretation by anyone who needs to extract information from the specification.

The reasons for the use of computer simulation as an aid to FMS planning are generally held to be as follows:

- The ability to operate a model of the system reduces the risk that the actual system will not operate in the intended manner.
- The ability to test alternative combinations of facilities and operating strategies enables a nearer optimal solution to be developed.

The justification is that the time and cost associated with model building and testing, although by no means small, is considerably less than the additional cost and lost time which would be incurred if an inoperable or sub-optimal system were installed.

As a result of a programme of work conducted as part of the Austin Rover/University of Warwick Teaching Company Scheme in which computer simulation is being used extensively in conjunction with a number of manufacturing projects, the authors believe that there are other benefits which arise from simulation in addition to its use as a pure decision-making tool. These may be summarised as simulation as an aid to improve:

- Awareness of the scope of the problem.
- Understanding of the complexities of the problem.
- Communications.

Also, the authors believe that the benefits vary depending on:

- Who writes the simulation model.
- What type of simulation package is used.

FMS at Austin Rover

It is not the purpose of this paper to describe the systems being developed and installed at Austin Rover, but rather to explain why simulation is seen as being so important in the planning stages. However, it is considered necessary to describe the type of systems involved and company policy on system design and responsibility for supply.

Whereas most FMSs are seen as the efficient alternative to small batch production, Austin Rover is seeking to apply the principles of flexible manufacturing to the production of parts in much higher volumes. These are traditionally handled by flow line methods with a great deal of highly dedicated automatic equipment, but this is no longer appropriate to satisfy a market which:

- Requires an increasing diversity of product specifications.

- Fluctuates in its demand for different product specifications.
- Constantly requires changes in product design incorporating the latest technology.

The use of FMS for medium high volume production in place of flow lines necessarily means that the systems are relatively large. The size of a system is in fact heavily dependent on the degree of flexibility that it provides; the greater the flexibility, the lower the productivity and hence the greater the number of machines required to meet output requirements. The degree of flexibility to be provided is therefore a major variable which determines the overall cost and complexity of the system and is in turn determined by company economic policy. In general, the greater the flexibility, the higher the capital cost but the lower the costs associated with changes which are quickly and cheaply made in software. The decision concerning the degree of flexibility that is economically justified can only be made with a knowledge of the configurations of alternative systems and hence their cost and also the full implications of changing an installed system.

Although Austin Rover has considerable experience in the design and construction of manufacturing systems, it was thought that a system involving large amounts of equipment which must all be compatible would be best handled by a single supplier who can ensure this. However, Austin Rover requires that all individual manufacturing systems meet certain criteria laid down by company policy and also have the compatibility to form part of the total computer-integrated manufacturing system. To achieve this it was necessary to define as closely as possible the specification for a system without overconstraining the potential supplier and thus preventing him from contributing ideas or technology to enhance the system.

The simulation package

The work carried out by the Austin Rover/University of Warwick Teaching Company Associates was primarily to examine the manufacturing implications of recent developments in flexible automation and to identify feasible manufacturing strategies to management. The task was not simply to simulate the ideas of others but to originate and test new system designs, and consequently the people involved were graduate production engineers and not operational researchers or computer scientists. A simulation package was therefore required which did not require an in-depth knowledge of the mathematics involved or extensive programming experience.

Initially ECSL was used, but a survey of the market at the time (1982) revealed that two systems, SEE-WHY[2] and FORSSIGHT[3], were available with colour graphics display and screen interactive facilities. The ability to produce an animated display of a proposed system was seen as being a very powerful way of inspiring confidence in management that a system did in fact behave in a particular way rather than relying on rather unintelligible printouts of graphs, histograms and tables. Eventually the SEE-WHY package running on a Cromemco Microcomputer with twin 8-inch floppy disk plus 1Mbyte hard disk was purchased on the justification that graphics were necessary to convince others of the findings. It was later appreciated that the graphics are a much more versatile communications aid.

The established role of simulation as a decision-making tool

Simulation was looked upon initially purely as a means of helping to make the decisions relating to the quantities of facilities required in the FMS and the rules governing the way in which the system would operate. The problem involved a considerable number of variables because initially neither the input to the system nor any details of the system itself were fixed. Simple models were constructed and used to determine the sensitivity of various systems to changes in product design and product mix. This provided details of the degree of internal flexibility that a system had to exhibit to deal with particular changes in input and still maintain a specified output.

At this level of modelling, decisions are made purely on the basis of the effect of the change on the system performance. That is, simulation is examining 'what happens if ...' and not 'why does it happen'. Nevertheless it provides basic data to help make decisions such as the compromise between productivity and flexibility, which has to be a pure management decision but is likely to be a better one on the basis of the data accumulated.

Simulation as an aid to awareness and understanding

After the type of manufacturing system which would meet the requirements for present and future flexibility had been established, it was necessary to consider it in much more detail so that a specification could be written to define exactly what was required.

Initially the scope of the problem was deliberately limited to consider only factors which were shown by the sensitivity analysis to be particularly important. Factors such as machine breakdowns, transport times, etc., were dealt with by simple average figures rather than by considering the complexities of dynamic interaction. The simplicity of the model meant that it was very easy to appreciate the problem and understand it, and the prime benefit of the simulation was the rapid and cost-effective way of obtaining numerical data. In this respect it did not matter who actually constructed the model and interacted with it to produce the results because there was little else to be gained.

However, as the model is developed to reflect more accurately the requirements of the real system, the very act of writing the program imposes an awareness of all aspects of the problem and by personally interacting with the model one acquires an understanding of 'why things happen' in response to the 'what happens if ...' questions.

Awareness is distinct from understanding since it is possible to understand fully what is perceived to be the problem when the real nature and size of the problem is not fully apparent. Simulation aids awareness of all aspects of the problem because to write a program mathematically defining and linking all activities requires discipline. When ideas are being formulated in the early stages of system design, it is very easy to take for granted that part of a system will operate in a particular way without the logic actually being defined. It is then possible to write a system specification which is imprecise or incomplete without this being known if there is no test of its coherency. However, in order to construct a computer simulation model, one is compelled to consider the consequences of all activities and to define logic controlling the activities, otherwise the program is unlikely to run. It is when de-bugging the program that awareness of factors not previously considered can increase dramatically.

The imposition of the discipline to make a program run does not eliminate the possibility of neglecting some activities entirely, but with the increased appreciation of the problem this is made less likely. Neither does the compulsion to define activities and logic necessarily guarantee that the assumed values and rules will be valid, but they can be tested and changed depending on their effect on the performance of the model.

It is also possible to ignore the discipline and remain unaware by 'cheating'. If a particular feature of a system is considered undesirable to simulate (e.g. on the grounds of complexity but relative unimport-ance or lack of data), it can be ignored and the program be made to run by inclusion of dummy entities and queues. In this case the model does not reflect the real world but the programmer is at least aware that the deficiency may result in lack of awareness of the whole problem.

Understanding differs from awareness in that it is possible to appreciate fully the existence of all factors affecting system perform-ance without understanding the complex interrelationships that can arise between them. Understanding develops as a result of repeated interaction with the model and is considerably aided by the use of graphics to display instantly the effects of changes in the model. Changes to variables such as quantities and times are relatively easy to achieve and their effects are usually fairly easy to understand from the pictorial representation of the system. Changes to the operating rules are invariably more difficult to make and often have considerable 'knock-on' effects which make the influence of the original change less easy to quantify and hence more difficult to understand.

The ideal model would be written to allow all variables, including logic, to be changed interactively via the graphics screen without the need to alter the source program. This is only possible if all possible changes that need to be made to the model are known in advance so that sub-routines can be written to cater for all eventualities. In effect this means producing a program which is capable of providing all the answers to questions which have not yet been asked. For anything but the simplest of systems, when simulation may not be warranted anyway, it is considered that this approach is totally impractical since it assumes complete understanding of the system, it requires considerable programming effort and it assumes that no other variables will be introduced.

The practical method of producing a model is to develop the program at the same rate as the development of the modellers' understanding of the system. In this way the model gradually evolves but the 'feel' for how the model responds to change can be used to limit the number and combination of variables which the model deals with. This makes the model simpler and less prone to program error but can be dangerous in eliminating options which may for various reasons be worthy of further consideration.

Who should produce the simulation model?

Whoever performs the simulation becomes the person who acquires the awareness and understanding of the problem. This experience is extremely valuable but is very difficult to convey to others.

If a problem can be well defined and the input data are known, the best people to perform a simulation study are an operational research group who are experienced in producing accurate, well-written models and can produce them quickly. Even if well defined and understood, the problem may be highly involved in computational terms due to the need to perform many calculations on a large amount of data, and sophisticated methods of data manipulation may be required to enable the model to run at high speed. In this case the sole purpose of the model is to produce quantitative results to help make decisions between options involving different variables. To the decision-makers, who are usually production managers, the origin and content of the simulation model is unimportant provided there is confidence that the model behaves exactly as the real system. Whether the modelling is performed in-house or by external consultants therefore depends on resource availability, previous knowledge and the possible value of experience gained for future use.

If a problem cannot be well defined and the input data is highly speculative, then a major benefit of simulation is to assist understanding and awareness of the problem to enable better definition. If data is lacking, assumptions must be made, and if the modeller is not familiar with the technical aspects and constraints of manufacturing there is a risk that wrong assumptions are built into the model, and these can be extremely difficult to identify later. System design may depend on the need to integrate with other systems and support functions and there is therefore a need for the modeller to incorporate any relevant requirements or constraints into the model. For this type of problem it can be argued that a wider knowledge of manufacturing than of simulation is required to produce the best model and that it is therefore easier to teach simulation to manufacturing engineers than it is to teach manufacturing to simulation specialists. This is reinforced by the recent development of computer simulation packages which are specifically designed for use by non-specialists within a relatively short time.

In using a simulation package there is a risk that an inexperienced user may not understand how the pre-written software routines handle data and it is quite possible to produce a model which processes correct data to produce incorrect results because it does not operate in the intended manner. As the level of sophistication or 'user-friendliness' increases it becomes easier to enter data correctly but in the attempt to reduce programming effort it is inevitable that the package makes assumptions which may not be applicable. Also, as the level of 'user-friendliness' increases the level of awareness and understanding declines because the modeller is partially removed from the model building process as the task is taken over by the package.

It may be thought that the best solution is to have a two-person team comprising a manufacturing engineer and a simulation specialist, but this can be considered to be the 'ultimate in user-friendliness' in which the simulation specialist is acting as an intelligent front end and output interpreter. This means that the manufacturing engineer is removed even further from the model building process and the discipline it imposes.

The model in this case is likely to produce invalid results not because the simulation package itself is incorrectly used but for the following reasons:

• The simulation specialist makes incorrect assumptions or misinterprets what the manufacturing engineer requires.
• The manufacturing engineer asks for the wrong data to be simulated because of his lack of awareness and understanding.

The conclusion reached as a result of this work is that, at least during the design phase of complex systems where simulation is being used to experiment and gain experience, the person who is best able to write the simulation model (i.e. the simulation specialist) is not the person to whom the experience is most beneficial (i.e. the manufacturing engineer). Consequently manufacturing engineers must be trained to use packages which require sufficient knowledge to use them to ensure they are used correctly but are sufficiently high level to eliminate the need to know exactly how they operate.

Simulation as an aid to communications

Simulation packages such as SEE-WHY and FORSSIGHT which have powerful colour graphics facilities can be programmed to produce an animated display of the manufacturing system (Fig. 1) and also to present performance data in graphical form which is quick and easy to assimilate (Fig. 2). The animated display in particular helps anyone not connected with the system design to understand the basic principles of

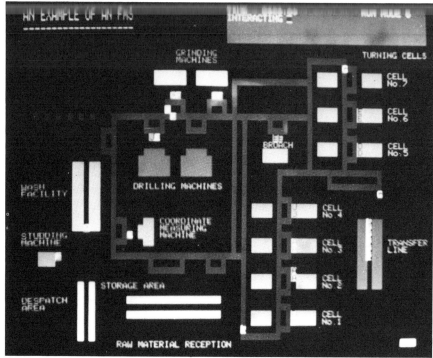

Fig. 1 Typical animated VDU display

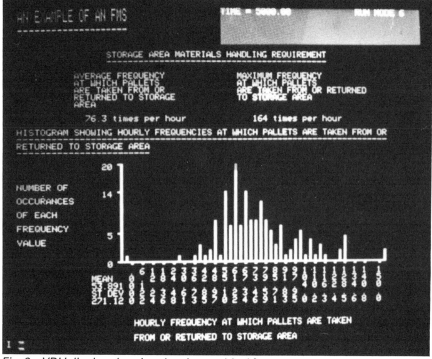

Fig. 2 VDU display showing data in graphical form

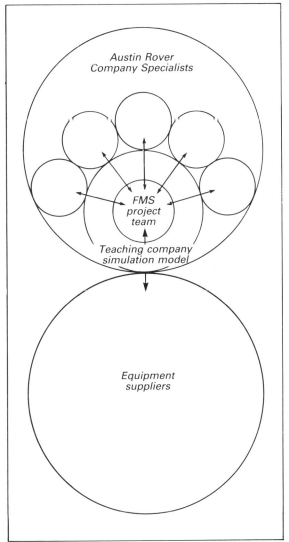

Fig. 3 Communications via the simulation model at Austin Rover

operation, and for those more deeply involved it can help to clarify specific aspects of the design. It has applications both within the user company and for external communications (Fig. 3).

Internal communications

The design of an FMS encompasses a wide range of disciplines both from the point of view of the facilities to be incorporated within the FMS itself and also with regard to the existing support systems with which it must interface. It is vital that the FMS project team communicates with the company specialists in all the related disciplines and that the graphical representation of the model allows system proposals to be quickly and effectively explained.

External communications

The potential suppliers of equipment for an FMS need a clear definition of the requirements for the system but should not be overconstrained. A written specification distilled from the base line simulation model is considered sufficient at this stage. When quotations and system proposals are received the most likely candidates' proposals can be simulated by adapting the base line mode to each of their proposals. This serves three major purposes:

- It ensures that the supplier has included sufficient information to enable the model to be built.
- It ensures that the user company appreciates how the proposed system is intended to operate.
- It ensures that the proposed system will meet the objectives.

In practice it is highly unlikely that a proposal will be adequately detailed or sufficiently clear or will have been sufficiently well tested to eliminate the need for further discussions with potential suppliers before a decision is made regarding final suppliers. The final choice is between a number of well-defined models the relative performance and costs of which are known, and hence the decision can be highly objective.

When a supplier is chosen the simulation model represents the system specification which is agreed between the two parties and can therefore be the basis for contractual agreement. It is, however, ensuring that all changes are understood and are acceptable to both sides, continuous updating of the model provides a common database which helps to eliminate misunderstanding and promotes smooth implementation of an agreed system.

Concluding remarks

The successful operation of complex manufacturing systems depends on good planning, which requires a total awareness of the scope of the problem and complete understanding of the facility interactions. Simulation provides an aid to good planning and enables precise definition of a system in terms of facility requirements and control logic. Planning effort is wasted if specifications are misinterpreted or misunderstood. The use of graphics facilities to provide animated displays of the simulated system helps to communicate ideas unambiguously.

To summarise, simulation is used in the design of FMS to assist:

- Decision-making:
  - Facility type.
  - Facility requirements.
  - Operating strategies.
- Awareness and understanding:
  - All the factors affecting system performance.
  - Effect of time-dependent interactions between facilities.

- Communication:
  - Between personnel within the FMS user company.
  - Between the FMS user and external companies or organisations.

Acknowledgements

The authors wish to acknowledge the SERC/DTI Teaching Company Scheme, which supports the programme between Austin Rover and the University of Warwick, and to thank Austin Rover for permission to publish this paper.

References

[1] Ingersoll Engineers 1982. *The FMS Report.* IFS (Publications) Ltd, Bedford, UK.
[2] Istel Ltd. *SEE-WHY Technical Brief.*
[3] Business Science Computing. *Forssight – Decision with Vision.*

5

Using Simulation for the Control of Manufacturing Facilities

The control function of a manufacturing system is an integral part of its design. This section concentrates on the use of the simulation technique to help in the process of designing or modifying control procedures for manufacturing systems. A simulation model of an existing plant or factory may be developed and then used to experiment with various control strategies. This same method may be used with an existing plant or factory and so becomes a powerful scheduling aid to management at both operational and strategic levels.

DESIGNING THE CONTROL OF AUTOMATED FACTORIES

R.B. Beadle
BLSL Inc., USA

The technique of visual interactive simulation is introduced. The advantage of this technique is demonstrated by examples that have been chosen to illustrate the particular relevance of visual interactive simulation to today's automated production units.

In the late 1970s BL Cars, British Leyland as it then was, was given support by the British government so that a replacement for the Mini could be built. This project was virtually a make or break situation for BL Cars. As a consequence of this situation every effort was made to ensure that the design of the manufacturing facilities was such that the design objectives for the new car could be achieved, and achieved in the shortest possible time. One part of this design study was the use of computer simulation to evaluate plant layouts. Included in the layouts were robotic lines, automated warehouses and computerised conveyor networks and it was in the simulation of these elements that there were initial problems.

These problems arose because of the need to integrate into the simulation control of these automated facilities. In the early stages of design no one knew what these control rules should be and when a change in the plant design occurred it implied a change in the control of the plant. To assist in solving these problems a physical model of proposed layout was built. This model consisted of a large wooden board and counters that were to be moved in accordance with the projected flow rates.

The physical model proved to be very useful in identifying weaknesses in the proposed facilities but it was an impractical means of evaluating solutions to these weaknesses. What was needed was a physical model that could be changed quickly to represent different designs and could be used in such a way that time could be simulated in accelerated mode. In other words a computerised visual interactive simulation model was needed.

At approximately this time some work was being carried out at the University of Warwick by Dr. R.D. Hurrion (the Editor of this book). Bob Hurrion had a prototype system that enabled a schematic diagram of plant to be displayed on a colour screen. Production could then be seen flowing through the schematic and, at any point, this process could be interrupted and modifications made either to the flow rates, the available facilities or the control of the production flow. These capabilities were exactly the features that had been requested by BL Cars when using the physical simulation model of the proposed plant.

Bob Hurrion's prototype system has provided the basis of the technique now referred to as visual interactive simulation. The capabilities of this technique are best illustrated by example.

A visual interactive model of a computer controlled conveyor network

A computerised conveyor network has been designed to carry car bodies from two robotic lines via a buffer to four manual finish lines. The purpose of simulating this system is two-fold: firstly to determine how many motorised slings are required in the network, and secondly to determine the rules for operating the system. The conveyor network was to be controlled by a programmable logic control (PLC) and it is the logic that will be applied by this machine that is to be determined by the simulation. There are several constraints on the system; namely, input rate of the buffer store must equal output of the buffer store at all times, car bodies must leave the store in exactly the same sequence as they entered it and the control rules must be capable of operating with any combination of load and unload stations operable.

A detailed model of this conveyor system was built such that all 72 conveyor stops were modelled and the time for a motorised sling to travel between two adjacent stops was modelled to within one hundreth of a minute. Fig. 1 was representative of the display that appears on the colour screen, with 'S' representing an empty sling and 'B' representing a sling carrying a car body.

The logic as designed by the suppliers of the conveyor network was then imposed on the model. The model was set in motion and the same suppliers were able to validate the model by checking both the logic of the flow and the timing of slings between various conveyor stops. They found the model to be an accurate representation of their design, and at this point the logic appeared to be meeting the design objectives. Then, using the interactive capabilities of the system, breakdowns of load and unload stations were imposed. As a result of these breakdowns the logic of the conveyor network failed causing loss of sequence through the buffer store. Under certain other conditions the simulation showed that the conveyor network would jam due to the fact that the logic of the system could not resolve competing demands for certain routes through the network.

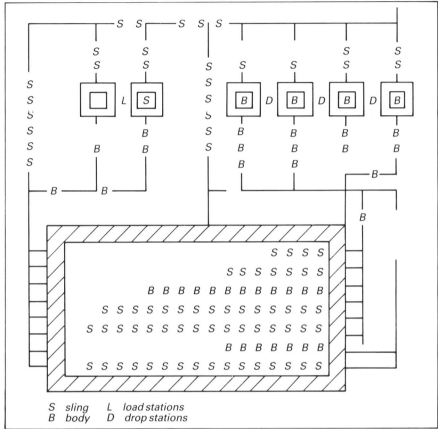

Fig. 1   Computerised conveyor network

Both of these situations could be seen on the simulation screen as they built up and as such it was obvious what the discrepancies and the errors in the logic were. Management were a part of these experiments, and they put the solutions to these problems to the test, such that a truly robust logic could be defined and accepted that enabled this system to meet its design objectives.

The complete study lasted approximately four weeks, involving one analyst for that time. The production loss that would have occurred if the logic had not been corrected was of the order of US $120,000 per week and the production build up of the new car would have been put back several weeks.

An automatic welding facility

An advanced engineering study had proposed that a system of automatic welding gates fed by automated guided vehicles (AGVs) should replace the conventional assembly line welding systems. Management were concerned about this approach on two accounts. Firstly, they could not easily size the facility requirements (how many

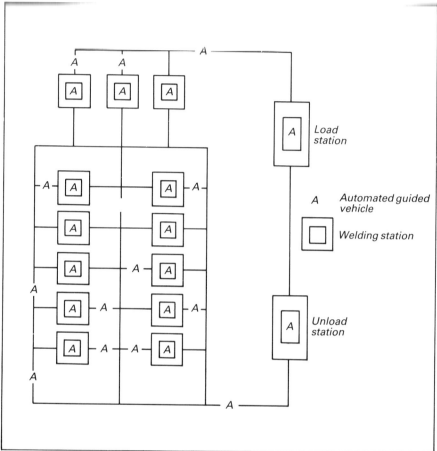

Fig. 2   Automatic guided vehicle system for welding car bodies

welding gates in what kind of configuration, fed by how many AGVs?), and secondly, how variable would the output from such a system be? If the output from the proposed system showed any significant variability then storage would be required to keep production flowing to subsequent operations.

A schematic of the proposed layout is shown in Fig. 2. Before this layout could be simulated, there was a need to determine how to route the AGVs through the six stages of manufacture. Initial constraints were that the AGVs must visit the primary weld stations first and then they could visit the five secondary weld stations in order.

Step one in modelling was then to put an initial set of routes into the model and observe the flows, under both normal operating conditions and under situations covering breakdown of weld stations. These experiments were carried out in the presence of design engineers and, therefore, the problem of communicating the results was eliminated. The results of this stage of the model were a set of control rules that depended upon the secondary weld stations being visited in sequence.

This was the only way, given the proposed layout and all feasible alternatives, to maintain a seemingly steady output from the system under all types of operating conditions.

The imposition of sequential welds on the system meant that some of the expected flexibility of the system could not be achieved in practice. This would have been very difficult to explain to the design engineer if they had not been able to participate in the experiments and view the flows through the system.

The second stage of the modelling was to determine how many AGVs were needed and how variable the output would be under expected operating conditions. These experiments were again carried out with management and design engineers present and again the results were surprising to them. The initial estimate of the number of AGVs required gave rise to a very variable output, and the reason for the variability was obvious from watching the dynamic display of the model. Whenever a breakdown occurred this caused the AGVs to bunch-up which alternatively starved and then overloaded the output stations. The immediate reactions to this problem was to increase the number of AGVs, this being easily achieved in front of management with the interactive capabilities of the system. The effect of increasing the number of AGVs was to increase the variability of the output and the bunching became more severe. Reducing the number of AGVs did reduce the variability of the output but it also reduced the maximum achievable level.

By observing all of these effects on the model in the space of one or two hours both management and the design engineers became fully aware of the operational characteristics of such a system. Obviously many different configurations were tried to track and machines but none of the configurations that were physically possible (there were space constraints on the layout) gave a suitable level and stability of output. Therefore, such an approach to manufacture for this situation was abandoned at a relatively early stage. Subsequent designs for this kind of manufacture have been much more realistic in their concepts, indicating that the design engineers have learnt a great deal from visual interactive simulation.

A flexible manufacturing system

Flexible manufacturing systems represent one of the largest changes in manufacturing procedures that are currently taking place. As with most radical changes many oversights occur that result in either under-achievement or in redundant elements. Visual interactive simulation offers many benefits in this area which has the sizing and control elements more closely aligned than any other manufacturing example.

The case study from this area is an eight machine system plus a wash and a load/reload area designed to produce 16 different components in vastly different volumes. Transportation within the system is provided by AGVs. The questions that were asked of the simulation were:

- How many AGVs are required?
- What routing and priorities should the AGVs follow?
- What effect does the sequence of manufacture have on the efficiency of the system?

The base design of the system had the machines aligned in two banks of four with a simple loop track for five AGVs providing the link between the machines. This design was simulated and was proven to be adequate in terms of the design objectives. However, the design engineers in watching the simulation observed queuing of the AGVs taking place at several points in the system. Consequently they asked that the simulation model be amended to reduce the queuing. Their suggestion was that a feed loop on the AGV track be included for each machine, the contention being that this would eliminate queuing at each machine but they wanted to know if these loops would reduce the total requirement for AGVs. This second design was simulated and as anticipated the queuing in front of each machine was eliminated but the number of AGVs required to meet demand varied between four and five dependent upon the sequence in which the various parts were made.

It was also apparent that the AGVs were travelling relatively long distances between operations. A reduction in travel between operations would be achieved if bypasses were introduced across the centre of the AGV loop. Again, the model was modified to include these

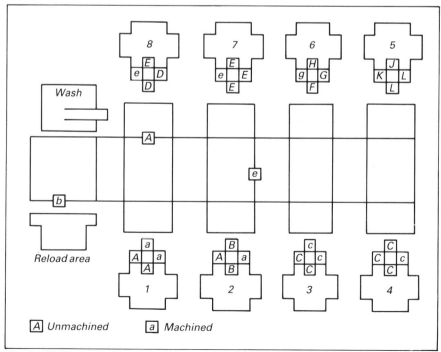

Fig. 3   FMS simulation

bypasses, but this time the effect on production was significant. Three AGVs could meet design production requirements independently of the sequence of production. The only cost being an increase in the complexity of the control of the AGVs which when compared to the cost of two AGVs was not significant. The final layout of this system is shown in Fig. 3.

Concluding remarks

Modern automated factories offer the design engineer many opportunities to use robotics and automated control. These facilities offer flexibility of manufacturing but they present a new type of problem. This problem is that the scheduling and the layout of the plant can no longer be considered separately. Visual interactive simulation is an excellent media for investigating these problems and for communicating the results. This latter ability to communicate results should not be underestimated. Given that every picture is worth a thousand words what is the value, in terms of written reports, of an animated picture that can be amended on your command?

THE USE OF SIMULATION IN CYCLE MANUFACTURING

T.M. Gough
TI Raleigh Ltd, UK

TI Raleigh uses simulation to support engineering project
developments and also to schedule such plant when it is implemented.
The intent is to plan properly plant investment and to operate this
efficiently post-implementation.

TI Raleigh has, over several years, been modernising its production
methods on its major site in Nottingham, UK. Those efforts have been
directed towards improvements in product quality, reduction in costs
and a greater responsiveness to the cycle markets of the UK and
Europe. These modernisation efforts have involved substantial
investments in new plant and management systems.

In October 1982 a decision was taken to employ simulation
techniques to assist in planning this investment programme. Subsequently a 'SEE-WHY' unit was purchased from the then BL Systems.
The intent was to model these intended investment schemes in order to
plan installations better and to understand operational characteristics
before plant was installed on site.

The last two to three years have seen a slow down in planned
investment. More effort has been concentrated on logistical problems
and simulation has been used to evolve plant scheduling methods and
manufacturing strategy. It should be understood that TI Raleigh's
product range is large and has true variation (some 750 unique models)
including toys, children's cycles, adult cycles and high-specification
lightweight cycles. While overall volumes are high, justifying large
investments in heavy process plants, other processes are light metal
cutting and manipulation, fabrication and assembly. The major issues
tend to be logistical. Thus the use of simulation is directed towards
these issues rather than to 'high-tech' solutions such as flexible
manufacturing systems (FMS).

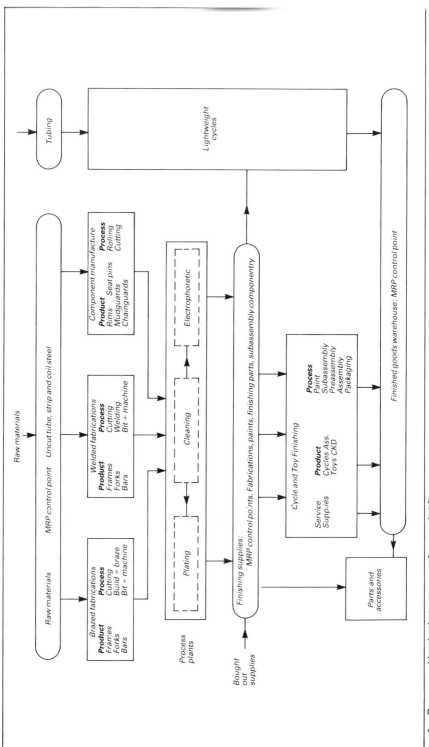

Fig. 1   Proposed logical structure of material flow

Engineering projects

The initial simulation project, undertaken with the assistance of Istel, was to model the storage of frames, forks, etc. (colloquially paint set items), before and after painting. This need was identified since the storage media was planned to be a power and free conveyor system. The cost of such installations meant that it was necessary to model beforehand to determine the size of the store. Size was reckoned a key factor in the feasibility of the store and its location.

The second major project undertaken was to model rim manufacture. A development and investment programme was to be started. Investments were to be made in the basic processes of rim rolling, welding, piercing and polishing. The task set was to model the number of units to be installed. The objective was to understand the required capacity balancing of these units and to model the operational characteristics.

Current work is centred on modelling the material flow of the company, illustrated in Fig. 1. As indicated earlier, the variety of products and the range of processes in use are large. These parts converge from their different routes on to final assembly. This process is planned by an MRP (material requirements planning) system. Degradation of performance in one channel of supply causes serious disruption throughout. This simulation exercise is designed to show the acceptable performance limits of these supply activities and the consequent inventory management effects. Armed with these facts, a major input can then be made to the design of the overall manufacturing structure, defining necessary performance limits, and thus influencing capacity considerations and store sizes.

Plant scheduling

The other major area of work in which TI Raleigh has been involved is work scheduling. As mentioned previously, the company employs an MRP system to plan at a *macro* level of work. Thus when a new £4 million paint plant was being commissioned no appropriate planning tools were available for work scheduling. Simulation techniques have been employed by a small multi-disciplinary project team to model this plant. Using a suite of programs, this work can now be reviewed prior to running it through the plant.

The plant consists of an electrophoretic coating primer plant, a store of primed stock, four finishing plants giving three methods of painting, and handspray cabins (see Fig. 2). All these features are, of course, embodied in the model.

The first part of the model defines the operating characteristics and capacities of these plants, rated in hangers of work (such a hanger may carry several parts). This data was collated by a simulation engineer working with a member of production supervision.

The second part of the model defines the product range in relation to paint colours and handspray, and relates those colours to the plants

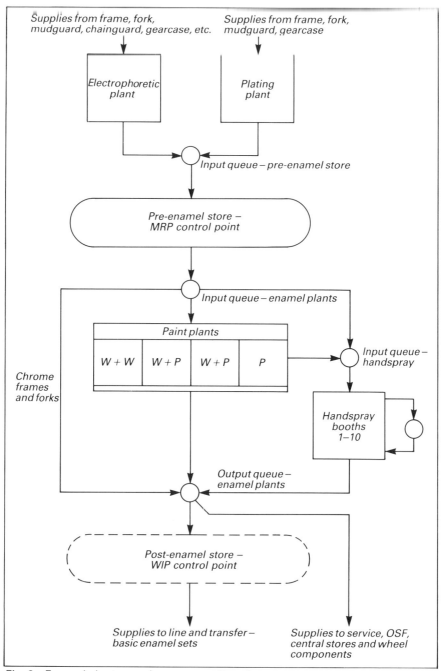

Fig. 2   Enamel plant complex: process flowchart

they run on. This data was collated with major input from the design function and production supervision.

The final section of the model gives the ability to input production programs, to run these through the model and thus to portray the impacts on the various paint plants. These input and reporting routines are separated into three modules:

- *Forecast review routine*: This shows a month's portion of a programme. It is portrayed as histograms showing plant utilisations. This allows first-cut load capacity planning. Its relevance is to both production planning to re-shape the load into an acceptable form, and to production supervision to pre plan plant operation and labour loading.

- *Order loading routine*: This shows a week's worth of orders. These are shown initially as segmentation to daily loads. This action is performed at the stage when orders are committed as firm to the manufacturing process. Key issues of balance of work and efficient use of plants are addressed by production planners and shop-floor supervision.

- *Daily order loading routine*: The process of scheduling work to daily blocks described above leads to shop supervision receiving a series of orders to process which has been rough-cut planned. The intention now is to sequence this work in order to ensure smooth flow through the shop and efficient plant organisation. The process of switching the sequence of work is depicted in terms of plant utilisation and the impact on handspray.

The use of the model is tied into the company production planning procedures. These procedures ultimately control the factory model program, the master production schedule (MPS).

This model program covers five forward months, including two months of firm orders planned to daily loads, and three months of forecasts planned to weekly loads. Typically the work routine is organised in the following manner.

The data for different plant characteristics (e.g. speeds, number of shifts being worked, process routes by paint types), are maintained by production supervision. The details of new cycle models (i.e. their colours and component content) are maintained by a production planner.

When the forecast section of forward factory schedule is being compiled each month, the production planner will 'run' the various forecast months and gauge their impact in load/capacity terms on the various plants. This is then discussed with production supervision as to its feasibility and desirability. The results are used to re-negotiate the program and conceivably to plan alterations to plant manning and operations. The intent is to ensure a feasible program which can be run in an economic manner.

The transfer of work from forecast to firm order status occurs weekly. The orders are inserted into a rolling eight-week horizon. They are now split into days of work and the model is used to assess this split. Plant loading is important not just to ensure overloading does not occur

212 SIMULATION IN MANUFACTURING

and jeopardise program feasibility, but also to plan the most efficient use of plant, minimising up-time and thus energy costs.

When the days of work thus planned are finally released into the factory, the final element of the model is used by production supervision to plan the detailed sequence of work. In this instance minimising change-overs is one key factor. Smoothing the flow of work to handspray cabins and thus preventing feast or famine situations is also of paramount importance.

The use of simulation in these models provides the means to evaluate rapidly different scheduling strategies. While the animation capability of the 'SEE-WHY' unit is not used, its strong graphic capabilities have been used to provide excellent, simple load/capacity data as a focus for the necessary debate between production planners and supervision.

A PRODUCTION CONTROL AID FOR MANAGERS OF MANUFACTURING PLANTS

P.W. Udo Graefe and A.W. Chan
National Research Council of Canada
and
M. Levi
Interfacing Technologies, Canada
(formerly of ICAM Technologies Corp., Canada)

An interactive computer model in a scheduling and modelling package being developed for small and medium-sized manufacturing plants is described. There are four main uses of the model. First, the model can simulate in detail the running of a given schedule to verify that assumptions made by the scheduler are valid. The scheduler and the model will interact to arrive at a feasible schedule. Secondly, the plant manager can run through tomorrow's schedule on the model and identify potential trouble spots. Thirdly, in actually executing today's schedule, the manager can use the model to respond to machine breakdowns and other unscheduled occurrences. Finally, the model can be used to evaluate 'what-if' type situations, where the impact of different facility layouts, manpower levels, number of shifts used, etc., can be investigated.

ICAM Technologies Corporation, in collaboration with the National Research Council of Canada, is currently in the process of developing FAMS, a Flexible Automated Manufacturing System software package, which will provide an integrated scheduling and production management tool for the small to medium-sized manufacturing job shop. Study of the Canadian manufacturing environment and analysis of some standard job shop operations and requirements lead to a conceptual definition of FAMS.

The general layout of FAMS is shown in Fig. 1. FAMS comprises a *kernel* and a *shell*. The kernel consists of three modules: The Scheduler, the Modeller and the Shop Monitor. The shell consists of two interfacing modules: the Database Interface and the User Interface.

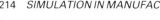

Fig. 1   General overview of the FAMS package

- *The Scheduler*: The Scheduling and Planning module produces detailed shop schedules as well as master production plans. User-selectable manufacturing goals, such as maximising throughput and/or machine utilisation, minimising work-in-process and direct cost, can be implemented in the optimisation process. Additional capabilities are job prioritising and on-line schedule modifications. The Scheduler, the heart of the FAMS, employs the Modeller's simulation tools to yield realistic schedules.

- *The Modeller*: This module assists the user in the decision-making process at different stages of the shop operations. Initially, when designing a new manufacturing shop facility, the Modeller is used to evaluate designed capacities versus projected demand profiles. In the same manner, any proposed modifications to an existing shop may be evaluated before actual implementation. At the day-to-day production management level, the Modeller is used to simulate schedules in order to predict the effect on shop operations, detect bottlenecks, critical hours, interferences, etc. The Scheduler utilises this module non-interactively to explore the validity of possible schedules. The user is able to play out 'what-if' scenarios and determine a desirable course of action based on the results.

- *The Shop Monitor*: Accurate feedback from the shop is essential for the successful operation of FAMS. The shop activity data are processed by the Monitor, which updates the FAMS Database. Exceptions can be reported to the system via the exception handler incorporated in the Monitor which updates all future activities accordingly. User status enquiries are answered with up-to-date information. The Monitor data are essential in statistical analyses and in establishing operation time and lead time estimates for better schedules.

The three modules are interfaced to the Database and the 'world' via the interfaces of the *shell*:

- *The User Interface*: This module consists of a menu-driven access system, terminal data entry posting and reporting as well as advanced graphics display routines. Status and statistical information is displayed in a graphical manner. The user is provided with a view of the shop layout depicting on-going activities such as machine status, queues at work centres, etc. Any selected job may be instantaneously traced on the graphics terminal. An extensive Help facility and a self-teaching package will enable the user to quickly become familiar with all FAMS functions. Using the self-teaching package the user will get assistance in the correct selection of different functionality options available within FAMS for efficient and correct FAMS implementation.

- *The Database Interface*: The Database access routines will enable access to the FAMS Database, as well as any existing user Databases. All information handling by FAMS programs will be done via the Database Interface routines. Application programs will be designed to be independent of the Database used. Database customisation will be included in this module.

This article will describe details of the Modeller and relevant parts of the User Interface.

Modelling methodology

The model representing the manufacturing activities is based on the interactive discrete-event simulation package ANEVENT[1,2], which was developed during the last 15 years by the National Research Council of Canada while collaborating with industry in the modelling and simulation of industrial plants[3-5].

For the purpose of direct schedule verification by the scheduling algorithm, the model can be run non-interactively in the asynchronous mode, where the model time will be updated instantaneously between successive events. The model may also be used in an interactive fashion by a model operator in the synchronous mode, where the passing of time between the execution of events will be controlled by a computer clock.

The ANEVENT package also provides the capability for the automatic creation of periodic snapshot files and the keeping of journal files, which allow the model user to return later to any desired instant of time previously simulated, in order to examine alternative schemes. This feature will be discussed in more detail below.

User interaction with the simulation model is effected at two levels. During the initialisation phase, the user interacts with the modeller via a set of menus, questionnaires and data input forms. Editing the databases corresponding to existing shop layouts, job profiles and shop statuses enables the user to modify the scenarios to be modelled. In addition, a command handler allows user interaction during interactive simulation sessions, when the modeller can output information to the user in three forms: graphical displays, CRT alphanumeric displays and printed reports.

An advanced full-colour graphics display package will be employed to reflect the status of the model at any point in time. Each display contains predefined dynamic fields which may contain numerical values, text strings, special symbols, graphs or histograms, which are periodically updated by status change routines driven by the model logic. Direct interface between the model programs and the graphics package permits real-time monitoring of manufacturing activities in the model. High-level picture creation facilities provide the flexibility for the customisation of these displays for each user. Reflection of the shop activities is effected using colour change, highlighting and animation techniques. Two examples of graphics display will be presented in more detail in the following section. Histograms and bar charts are used to report relevant statistics, such as status of schedules, machine loading, material handler utilisation and queue levels. A sequence chart may be turned on by the user to provide a hard copy of the model status at predefined time intervals. During interactive runs, the user may interrupt the model at any moment. The command handler is automatically activated whereby the user interacts with the model by issuing commands to request output reports and statistics as well as to control machine schedules, equipment breakdowns and resource utilisation.

Model description

Input to the model can be thought of as comprising two main components: a shop image and a job profile. The shop image in turn is subdivided into a static and a dynamic part, while the job profile consists of a time-dependent schedule and a job-dependent bill of operations.

The static shop image describes the shop layout – that is, the location of all machines, material handlers and their paths, and storage areas. It also specifies the characteristics describing these machines, material handlers and storage areas. Shop operating characteristics such as the start times and durations of the various shifts and break periods also form part of this input.

The dynamic part of the shop image provides information on the current status of all machines, material handlers and storage areas, including a description of parts which are currently in the manufacturing process. A description of the currently available human resources, their skill levels and their status is also included.

Time-dependent data for the job profile are given by the schedule, which provides the number of parts to be made for each job and the start and finish times for each scheduled triplet of job-operation-machine. Apart from regular machining opeations, the model includes such operations as set-up, tear down and NC program proofing. Since the set-up and teardown times are sequence-dependent, the times required for these operations are also provided by the scheduler to the model.

The job-dependent bill of operations specifies the routing through the shop to be used by the parts of a job during manufacture. It consists of a sequential list of operations, including material handling operations between machines and storage areas. For each machining operation the machine and machining time as well as the probability of the part having to be scrapped is given. For material handling operations the specific handler (if dedicated) or handler type is provided. The number of units to be moved at one time is also specified for material handling operations, allowing material transfers of individual pieces as well as of batches with each move.

There are four methods of initialising the model. First, for a new shop a cold start requiring all input data would be used. Secondly, for a shop that has been modelled previously, the static shop image is not required. Initialisation from data provided by the shop monitor would be appropriate in this case. While an interactive simulation run is in progress, the model will automatically produce periodic snapshots of the entire model database, as well as keep journal files of all user interaction with the model. This provides the basis for the third and fourth methods of initialisation. To initialise the model to correspond to the state at which a previously simulated run has been terminated, only the relevant last snapshot and a new schedule and any necessary bills of operations are required as input. To return to a given shop situation in a previoius simulation run, one merely has to identify the previous case simulated and the desired time. The model will then be initialised from the appropriate snapshot and journal files.

After initialisation, the model is ready to run. At the start of a new shift, or after the completion of a job operation on a machine, the schedule is examined for one machine at a time, to see if there are further job operations such as NC program proofing, set-up, machining or teardown to be performed at this time or at some time in the future. If a new operation is scheduled to start, all required conditions and resources are checked. These include a check on machine status (e.g. is it set up for the next operation?), availability of an operator or set-up person with the proper skill level, and the availability of a part. If all requirements are met, the task to load the new part of the machine is issued to the appropriate material handler or material handler group. If none of these handlers is available at that time, the task is entered in a 'look for work' list from which tasks are later dispensed, one at a time, as appropriate material handlers become available.

After a part has been loaded on a machine, the machine will process the part for the duration specified for this part and operation in the bill of operations. At the end of the operation, a random number based on a specified statistical distribution for scrapping of parts for this operation is generated to determine if the current part is to be scrapped. If the part just produced meets specifications, the bill of operations is examined to determine the next location to which this part is to be transferred, and the appropriate material handler will be charged with this transfer task. Conveyors as material handlers are treated somewhat differently in that they can accept parts to be moved at any time as long as there is room on the conveyor and the conveyor is operational.

Details reflecting numerous common occurrences in a machine shop are accommodated in the model. Some machines in the shop may be capable of processing more than one part at a time, such as multi-spindle machines. In this case, the model will allow multiple loadings before machining starts. Other machines, referred to as batch machines, such as heat treatment furnaces, may accept a large number of parts before starting their operation, and their operation period may extend into or bridge time periods during which there is no working shift in progress. Human resources are allowed break periods during which some machines may be shut down. If one shift is not immediately followed by another, the model operator may elect to have some machines and human resources work overtime in order to catch up on a slipping schedule. Machines and material handlers may experience breakdowns or delays which will necessitate the use of alternative machines for some operations. To accommodate last minute changes, the operator can issue model commands to insert rush jobs, cancel jobs or rearrange jobs during the simulation. The model also provides cost reporting. Costs are computed for each job based on machine run costs, machine overhead costs, labour costs and a one-time charge per job for such items as programming, administrative costs, etc. Inflated rates apply to overtime periods.

The output of the model falls into two categories: status reports, which are associated with a specific point in time, and statistical reports, which summarise important information over a specified time period. The output may be in the form of printed reports or displays on a graphics terminal, the latter being very important for supplying the model operator with relevant status information needed for effective interaction with the model.

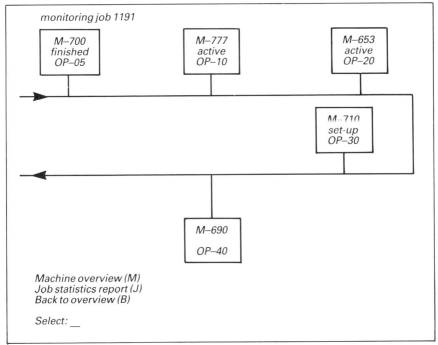

Fig. 2   Shop overview – job monitoring

Figs. 2 and 3 illustrate two typical displays on a graphics terminal. Suppose that the model user is simulating a job shop's operation on November 9, 1984. Specifically he is interested in job 1191. Selecting the Shop Overview-Job Monitoring display (Fig. 2) for this job will provide him with a view of all machines in the shop, with all the machines associated with job 1191 highlighted in a specific colour. In this example the model operator will see that the job requires operations on all five machines in the shop. Operation 5 on machine 700 is finished while operation 10 on machine 777 and operation 20 on 653 are still in progress. Operation 30 on 710 has not been started yet, although the machine has been set up already. Machine 690 is to perform operation 40 later, but is not yet actively involved with job 1191 – it may in fact be processing another job now. Based on this information the model user may now wish to obtain more detail on one of the machines involved, say machine 777. Selecting the 'Machine Overview' item from the menu displayed near the bottom of the screen or through the use of a joystick, he will be given a new display as shown in Fig. 3. This shows in graphical form that the machine is a four-axis machine, actively working on operation 10 of job 1191. Upstream of this machine is a job queue containing components for five separate jobs. Material handler 23 loads parts into this job queue while material handler 39 loads parts from the queue onto the machine. Parts are removed manually from the machine to an output storage area from where they will be removed later by material handler 17. The output

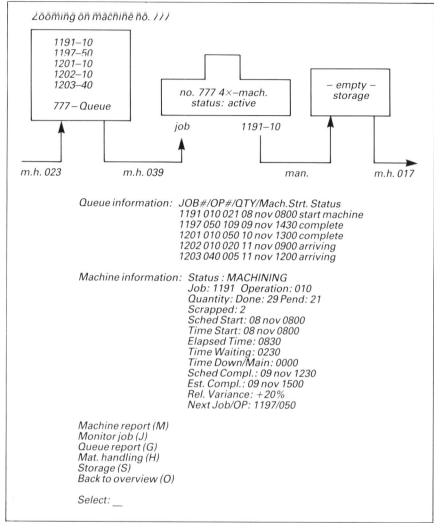

Zooming on machine no. 777

1191–10
1197–50
1201–10
1202–10
1203–40

777 – Queue

no. 777 4×–mach.
status: active

– empty –
storage

job   1191–10

m.h. 023   m.h. 039   man.   m.h. 017

Queue information: JOB#/OP#/QTY/Mach.Strt. Status
1191 010 021 08 nov 0800 start machine
1197 050 109 09 nov 1430 complete
1201 010 050 10 nov 1300 complete
1202 010 020 11 nov 0900 arriving
1203 040 005 11 nov 1200 arriving

Machine information: Status : MACHINING
Job: 1191 Operation: 010
Quantity: Done: 29 Pend: 21
Scrapped: 2
Sched Start: 08 nov 0800
Time Start: 08 nov 0800
Elapsed Time: 0830
Time Waiting: 0230
Time Down/Main: 0000
Sched Compl.: 09 nov 1230
Est. Compl.: 09 nov 1500
Rel. Variance: +20%
Next Job/OP: 1197/050

Machine report (M)
Monitor job (J)
Queue report (G)
Mat. handling (H)
Storage (S)
Back to overview (O)

Select: __

Fig. 3   Machine overview

storage area is currently empty. While a material handler is performing a transfer of parts, the corresponding line representing the transfer will change colour, indicating this activity.

Textual information on the input queue and the machine is given below the graphical display. From this it is apparent that the queue still holds 21 parts for job 1191, which has been processing on the machine since 8 a.m. on November 8. All parts constituting jobs 1197 and 1201 are present and complete in the queue, while more parts for jobs 1201 and 1203 are still arriving. So far, 29 parts for job 1191 have been completed, and two have been scrapped. The current job started as scheduled, but encountered a delay of two and a half hours, which means that the estimated completion time will be at 15:00 hours on

November 9, rather than at 12:30, as scheduled, giving rise to a +20% variance in schedule. The task to be performed thereafter will be operation 50 of job 1197. The menu displayed at the bottom of the screen allows the operator to select further displays for a variety of model components.

Model uco

With the most significant events in the day-to-day operation incorporated in the model, there are numerous ways the plant manager or the production planner can use the model. Four main applications of the model are presented.

Validation of schedules

Schedules being generated by the Scheduling Module of FAMS are sent to the Modeller to be validated as part of the scheduling process. During the computation-intensive process of developing the schedule, the Scheduler utilises global average parameters assigned to jobs and operations in the process plan, such as standard lead times, average waiting times for material handlers, and interprocess delay times. In addition to meeting user-defined criteria, accountability for the real job shop events will help to ensure a realistic schedule.

The Modeller can simulate in detail the activities in a job shop on a part-by-part basis. An example is the interferences between shared resources such as material handlers. Delays caused by such interferences are being taken into account during the validation process. A non-anticipated requirement of some material handler by more than one job at the same time may cause a significant delay, which may or may not result in overall schedule slippage.

Model initialisation in this case is a relatively simple procedure. The shop static image is the actual shop definition which exists in the Database. The status of the shop is given by the Shop Monitor, and the schedule is provided by the Scheduler along with some run parameters such as the time limits of the simulation. User interaction is not required and the whole validation process is carried out in the fast asynchronous mode.

Simulation of future schedules

Once a schedule has been generated, the user might want to step through it in a 'speeded-up' manner in order to observe shop activities before actual schedule implementation, to detect critical and slack periods throughout the shift or day, and to establish the human resource level and capacities required. Potential benefits include the utilisation of any time gaps to execute odd jobs or 'soft' production jobs, the forewarning of critical periods when extra attention or outside help is warranted, and the anticipation of slack periods which can be selected to relieve resources if required. The model user can 'replay' critical periods which had been observed by stepping back to any previously simulated moment using the snapshot and journalling capabilities. Model intialisation is again relatively straightforward. Actual shop definition, status and schedule information all reside in the Database, and user interaction is minimal during initialisation. The model will be used in the synchronous mode, where the user may select any desired speed-up factor and full interrupt capability will be available.

Simulate exceptions and unscheduled events

In the course of actually executing the schedule in the shop, unscheduled events often occur, some of which are significant enough to cause perturbations in the schedule. Some examples are machine breakdowns, unscheduled maintenance, job cancellation, change in engineering specifications, order for rush jobs, etc. In order to overcome any unscheduled disruptions which have arisen, the user may want to explore a number of alternatives. By means of the snapshots and journal files, he can try different options starting from exactly the same point during the simulation run. Via the available model commands, he can implement corrective actions and modify the schedule. The model then runs through the modified schedule and reports on any impact the corrective actions may have. The user is thus provided with a means of comparing and evaluating different alternatives when faced with unscheduled occurrences. Intitialisation in this case is a short procedure and the user can rely on snapshots and journalling files generated during previous runs in either use-type 1 (validation) or 2 (future schedule simulation).

Simulate 'what-if' scenarios

The 'what-if' simulator is an essential tool for many decision-making applications. It allows the effects of any proposed changes to be viewed before the actual implementation. For example, the impact on machine schedules and productivity goals of adding a machine can be observed by simply redefining the shop image and simulating a known job profile with the new resource complement.

Correct model initialisation is the key for a successful 'what-if' session. There are three elements in the initialisation process: shop image, shop status, and schedule (or a job profile).

Shop image initialisation is required when the 'what-if' simulation is concerned with any changes of the shop layout, number of machines, material handlers, etc. A previously defined image (the actual one or a hypothetical one) may be slightly modified. Different shop images can be tested during initial shop design or whenever a modification is contemplated.

Shop status initialisation is needed whenever a 'what-if' experiment is performed on a schedule. Status information may be obtained from the shop monitor or a previously created snapshot. As the status of a shop must be compatible with a specified image, then if the user had modified an existing shop image, the status image would have to be adjusted accordingly to maintain compatibility. A manually created status or a cold start status can also be entered by the user.

Few options are available to the user in initialising the schedule or the job profile to be simulated. A schedule must be compatible with the shop image and status. If a new or modified shop is simulated, a job profile may be chosen and the scheduler must be called to produce the schedule for the specified shop.

Concluding remarks

Phase 1 of the Scheduler and Modeller are now being used on a trial basis at a mould and die shop of International Tools Inc. in Windsor, Ontario, Canada. Based on the experience and feedback gained from this test, an improved version will be developed during phase 2 and tested at several different job shops before marketing of the product by ICAM Technologies Corporation.

While several other commercial packages[6,7] are available to perform capacity planning and scheduling, the FAMS package has the additional capability to schedule and monitor shop acitivies in great detail while taking into account material handling bottlenecks, breakdowns and other unscheduled occurrences. The package also provides the plant manager or other users of the system with interactive modelling capabilities to alert them to critical situations and to enable them to evaluate recovery strategies after upsets have occurred in the shop.

References

[1] Graefe, P.W.Udo. 1982. ANEVENT – An Interactive Computer Modeling Package Based on Discrete Event Simulation. National Research Council of Canada Report LTR-AN-48.
[2] Crate, G.F. 1982. ANEVENT – An Interactive Computer Modeling Package Based on Discrete Event Simulation, A Users' Guide. National Research Council of Canada Report LTR-AN-53.
[3] Nenonen, L.K., Graefe, P.W.Udo. and Chan, A.W. 1984. Industrial applications of interactive computer models: A decade of experience. In, *18th APCOM Symp.*, 26-30 March 1984, London, pp. 713-722.
[4] Graefe, P.W.Udo. and Nenonen, L.K. 1975. An interactive computer modeling approach to the study of materials processing and handling in the mineral industry. In, *13th APCOM Symp.*, 6-11 October 1975, Clausthal, West Germany, pp. 5-II-1-16.
[5] Graefe, P.W.Udo., Nenonen, L.K. and Strobele, K. 1974. Simulation of combined discrete and continuous systems on a hybrid computer. *SIMULATION*, May 1974.
[6] PERA 1983. Production Systems – The PERA 4 W Capacity Planning and Scheduling System – A Standard and Portable Package. PERA Report 379.
[7] PERA 1982. Production Systems – Capacity Planning and Scheduling Applied to a Group Technology Manufacturing System. PERA Report 372.

6

Simulation as a Routine Management Practice – Simulation Standards

The use of discrete-event simulation methods to assist with management decisions is becoming routine practice for many manufacturing organisations. This section considers the problem of simulation standards and the papers describe how simulation strategies have been adopted and evolved within organisations. Who should initiate a simulation feasibility study? Who should undertake the simulation project? Engineering staff, internal or external consultants? What are the key points in managing a simulation project? These are some of the points addressed.

SIMULATION ON MICROCOMPUTERS THE DEVELOPMENT OF A VISUAL INTERACTIVE MODELLING STRATEGY

R.W. Hawkins, J.B. Macintosh and C.J. Shepherd
Ford of Europe Inc., UK

The development of a strategy for implementing visual interactive modelling in Ford of Europe is described. The importance of making the simulation process readily available to the user engineer is shown to be the cornerstone of the approach. The problems of developing large-scale visual interactive simulations on microcomputers within this environment are discussed, together with the methods adopted by the authors for overcoming them. The benefits of modelling in this manner are also given and two case studies presented.

The Operational Research Department of Ford of Europe did not begin to use simulation modelling techniques in earnest until 1977. This date coincides with the planning of a new Engine Plant at Bridgend in South Wales. Simulation models of the proposed cylinder head and cylinder block machining lines were developed. These early models were so successful that requests were received to model a number of existing facilities in order that a better understanding of their operation could be gained.

At first simulation models were developed using either GPSS or Fortran in a batch environment on large IBM mainframes. Since January 1982, with an ever expanding workload and conscious of the need to develop models more effectively, a different modelling approach has been adopted.

The approach developed is based on the use of simulation models which are visual, interactive, and run on microcomputers. By adopting this method it has become possible to take the models directly to the project sponsors. In the comfort of their own office, users can see the models working and interact with them by directly changing many variables easily and quickly. Because of this involvement the users are more committed to simulation, they more readily accept results generated by the models, and, in many cases, are starting to build their own models.

Simulation modelling projects now cover the whole spectrum of Ford's manufacturing operations in Europe. Models are being used to assist in the planning of flexible manufacturing systems (FMS) as well as the more traditional automation normally associated with the motor industry. Where changes are being planned to existing facilities or where problems with existing facilities are being experienced, then simulation models are providing valuable information. With the advent of monitoring systems on the shop-floor the opportunity for the development of simulation models for crisis management is also being investigated.

The increasing complexity of most models which are now being developed and the limitations of memory and speed of processing on microcomputers has made it necessary to develop methods which will help overcome these problems. The use of these methods, described in more detail later, is now an accepted part of building visual interactive models in Ford of Europe.

Introduction to visual interactive modelling

The essence of visual interactive modelling is the ability to display on a colour screen a mimic diagram which the model user can relate to and receive as a moving picture of the system being studied. In addition, the user can stop the model at any stage to check, in detail, its workings and modify any element of the model. Thus, decisions are made by the user who is helped in making his choice by looking at the current status of the problem area displayed on the screen.

There are many advantages in using the visual interactive approach to simulation and some of these follow directly from the two principles defined above. Using a moving colour picture as output ensures that everyone involved in the study completely understands the rules used and the assumptions made by the model. The model is easier to validate and this must increase the confidence in the model of both the user and the model builder. Also, by presenting an overall and structured view of the study area, counter-intuitive patterns of behaviour can be seen and this new information can lead to better solutions to problems. The extensive interactive facilities within these models allow the user far greater control over the model. The user can interactively change the layout on the screen, the model data, or the control rules embedded in the model at any time. Furthermore, the model can be programmed to stop when certain conditions are met and prompt the user to make a decision. When the decision has been entered the model continues to run and demonstrates to the user the possible consequences of the decision taken.

A full discussion of visual interactive modelling and further references can be found elsewhere. In addition, an extensive review of the use of visual interactive models has been carried out[4].

The implementation strategy

At the same time as these advances were being made in visual interactive modelling rapid changes were also occurring in the technology of car manufacture. New types of machines were becoming available and new ideas on how to operate them were being developed. Ford of Europe was committed to using the new technology and required methods of evaluating proposals in this increasingly complex world. Simulation had proved an effective method to plan new facilities in the past and it was anticipated that visual interactive modelling would be an even more powerful tool. Thus, the new modelling software matched a real business need in the company and it was decided to promote the use of simulation within Ford of Europe.

The Operational Research Department was asked to develop a strategy for implementing visual interactive models within the company. Two key decisions were quickly made. For the first part, Operational Research would take charge of coordinating the implementation of visual interactive modelling software. Secondly, every attempt would be made to involve the user in the simulation process with the ultimate aim that they would become model builders as well as model users.

To begin with, models were developed by Operational Research that were simple to operate, easy to understand and could be changed without reprogramming. The users were encouraged to take responsibility for the model by borrowing a microcomputer and using the model at their own desk. For the first time within Ford of Europe the user could validate and experiment on a simulation model. This policy proved very successful and demand for the loan microcomputer quickly became so great that individual departments bought their own systems. There are currently a total of 19 visual interactive modelling systems installed in Ford of Europe.

The effect of the implemention strategy on users

The use of visual interactive simulation models on microcomputers located in the user's work area has led to a number of significant changes in the way simulation models are used and in the time spent on associated activites.

In order to obtain the full benefit that a model has to offer it was found necessary to train users in the practicalities of setting up and running a completed model, and the use of the host simulation software on its associated micro equipment. To this end internal training courses and refresher courses were developed and attended by the major users of visual interactive models in Ford of Europe. The courses covered the basic principles of discrete event simulation, the model interactions available, and the statistical methods required to analyse results.

Training for model builders was also found to be essential. Therefore, those members of the Operational Research Department whose background had not covered simulation attended courses on simulation as well as software familiarisation. Similary, any potential developer in a user area was strongly recommended to attend simulation, programming and software courses rather than rely on second-hand information or trial and error.

In addition, a study group has been set up to exchange information amongst model users. This ensures that everyone benefits from the modelling experience being built up in the many user areas and that Operational Research keep in close contact with the users. Here, for the first time, people from very different functions within the company regularly meet to discuss common simulation problems. This has been a unique development directly resulting from the strategy of implementing visual interaction modelling in the user areas and is considered to be a major benefit to all involved.

At the study group meetings it has been possible to set up sensible modelling standards, keep a record of work-in-progress to avoid duplication of effort, and ensure that everyone is aware of the latest development in simulation methods. The meetings are hosted by different areas in turn and may be preceded by a workshop where models under development are brought along and run on local equipment, viewed and discussed. Subcommittees covering an aspect of simulation common to more than one area were set up and they regularly report back to the study group. The modelling of conveyors and machine breakdowns are two examples of such common areas of interest.

Study group meetings are attended by both users and model developers; and by industrial engineers who may be both. In general, complex simulation models are programmed by Operational Research and simpler models by industrial engineers. Recently, a new type of model has been developed for the users who do not have programming skills. These 'generic' models are flexible enough to build significantly different models and the user can interactively build a model through the use of menus and command lines. This approach has made model development easier and engineering users can now contribute sophisticated skills in model building to the study group.

The success of visual interactive simulation in user areas can be measured by the number of people Ford of Europe has trained to use simulation models. There are currently over 60 engineers trained in the use of models and 19 microcomputers dedicated to visual interactive modelling in Ford of Europe. The Operational Research Department has three to four people developing models and there are three industrial engineers who spend a large amount of their time programming their own models. The simulation study group meetings are attended regularly by 25 people who take a full day out of a busy area of the company to contribute and exchange information. These figures are considerably higher than the maximum number of people ever involved in simulation using batch orientated software.

The model building process

This section discusses the effect of the implementation strategy on the model builder under the headings of analysis, model structure, and design. It would be beyond the scope of this paper to give a complete description of the model building process; rather, this section concentrates on some of the methods that have been developed as a result of using visual interactive modelling on microcomputers.

Analysis

An initial feasibility study is undertaken to ensure that visual interactive modelling is an appropriate tool for investigating the problem area. At the end of the study the analyst should be able to state the objectives of the investigation and understand how a simulation model will help the user to solve their problem. A detailed report is then prepared describing the rules that govern the system and the data that is available. The report is usually written by the user under the guidance of the model builder. Thus, the user is involved at an early stage in the modelling process and this promotes their committment as well as ensuring their cooperation in the project.

On many occasions it has been found that data is readily available but very little understanding of the control rules exists. Interactive models give the inexperienced user the ability to change data easily, therefore less time need to spent collecting data during analysis and the model builder can concentrate on exploring the logic of the system under study. It is interesting to note that a simple visual model may be developed at this early stage, as it has been found that even a static display on a colour monitor can help communication between model builder and user.

The final stage of analysis is to structure the model. A variety of formal techniques is used within the department and these include entity life-cycle diagrams, activity cycle diagrams, and state-event diagrams. All of the methods require the analyst to draw on paper the logic of the problem area using a simple set of symbols as building blocks. The diagrams are then presented to the user who checks that they correctly describe the system.

Model structure

The first releases of visual interactive modelling software bought by Ford of Europe ran on a Cromemco Z80A based 8-bit microcomputer. Through the use of memory bank switching techniques 192Kbytes of random access memory (RAM) were available on this system. However, when the operating system and host simulation software were loaded there was only 34Kbytes of RAM available to code the model logic. Fortunately, the storage of the simulation data array (information on entities, sets, time series, histograms, etc.) was an overhead on memory occupied by the host simulation software.

It soon became apparent, with the level of detail being modelled and the size of the problems studied, that the 34Kbytes of RAM were going to be a severe limitation. Methods of structuring models to overcome this restriction were developed, and this section descibes those methods found most successful.

All simulation models developed require a definition phase which defines and initialises the simulation data array. The code for the definition phase is usually only executed once during a simulation run and is an unnecessary overhead if it continues to occupy part of the 34Kbytes of RAM during the rest of the simulation run. With large models the code for the define phase could occupy as much as 20Kbytes of RAM.

A method of overcoming this restriction was developed based on separating the define and execute phases into two programs known as the define and execute models. The data interface between the two models was the define model's simulation data array stored externally from the model on the microcomputer's hard disk. Thus, using

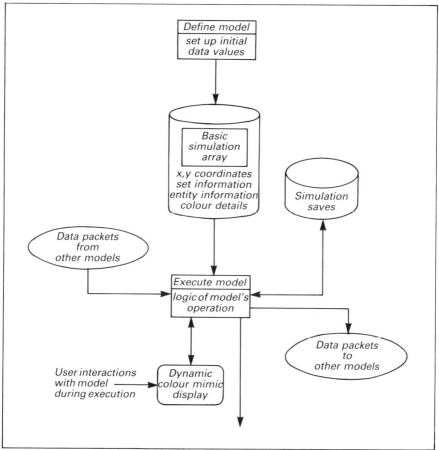

Fig. 1   Two-stage model

standard model interactions the initial simulation array would be saved onto the hard disk from within the define model. Subsequently, the execute model would then be run and the previously saved simulation array restored into the model. This method removed the overhead associated with the define phase and allowed the analyst to program larger execute models. It is interesting to note that the save and restore interactions of the host software are also used extensively to save states of the execute model after various experiments and at different simulation times. This facility is considered invaluable when carrying out a series of experiments.

Fig. 1 illustrates the concept of define and execute models. The diagram also shows data packets entering and leaving the execute models. These data packets act as information flows between a number of micro models and a macro model which are used to describe fully the plant or process being modelled. Data packets are a feature of very large model developments when even the two-stage model approach does not yield sufficient memory.

Intially data packets consisted only of the absolute clock time that entities left the micro model. The list of times provided an arrival sequence for the next micro model in the process. This approach proved unsatisfactory when entity type information was also called for in the successive micro model. The data packets were improved to contain coded information on the entity type and associated attributes as well as absolute clock time.

At the macro level, arrival times may be taken directly from the data packets or they are combined to produce arrival distributions which are then sampled from on a random basis. Also at the macro level data packets are produced which can be used by lower level micro models.

Use of micro and macro models in this way obviously does generate a number of difficulties. Whilst it is possible to model forward interference accurately it is not easy to model backward interference successfully. To overcome this problem the user is provided with sufficient information to enable him to test the preceding models in the process. Improved methods for linking models are under investigation.

An example of the use of micro and macro models is shown in Fig. 2. The Bridgend Engine Plant is again featured but this time the engine assembly lines are being modelled. Modelling to this degree represents an investment of approximately 7000 lines of executable Fortran code without taking into account the host simulation software overhead.

Since the beginning of 1984 visual interactive simulation models have been developed using 16-bit microcomputers based on the Motorola 68000 processor. The Cromemco microcomputers were field upgraded to 16-bit mode and 512Kbytes of RAM added to each machine. The new processor and additional memory did alleviate some of the difficulties brought about by model size and speed limitations. However, during the conversion of existing models it soon became clear that the method of two stage micro and macro modelling still had

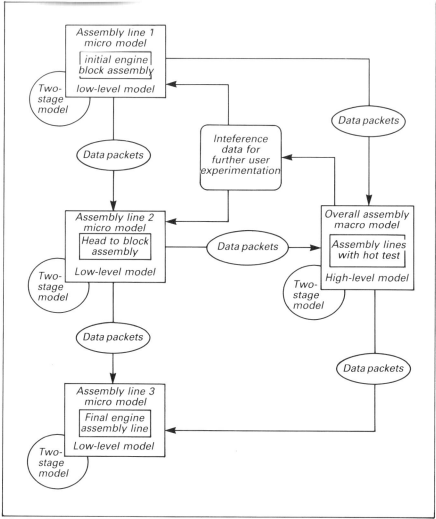

Fig. 2   CVH engine assembly model Bridgend Engine Plant, South Wales

an important role to play in the improved computing environment. In particular, the CVH engine assembly model was still too large to be programmed as a single model within the available memory, and, even if memory size had been increased, it was likely that the speed of execution would have been too slow to allow useful experimentation.

Design

The use of visual interactive software that offers a wide range of graphics facilities has led to more time being spent on the screen design of a model. The power of a pictorial representation of a study area has encouraged the model builder to design screens that clearly depict the facility being studied. Shapes, colour, and movement are used to highlight critical situations as they develop and to present statistical output from the model.

Large detailed models often contain too much information to fit on one screen and, rather than buy additional hardware, a method of paging has been developed to overcome the problem. Typically a schematic macro screen giving an overview of the model and a number of more detailed micro screens, which may overlap, are used to display the model's changing status. Each screen display is known as a page and there may be as many as 20 different pages for a large model. The user can easily select the page to view while the model is running by pressing one of the alphanumeric keys. The screen is cleared and filled with the new page representing the current status of the chosen area. Paging is considered to have great potential for further development and, ultimately, a large number of screen pages are possible for one model.

Several advantages arise from the use of pages. For example, having a number of different pages showing the status of a particularly critical area of a simulation model greatly improves the understanding of the problem. In addition, the visual interactive models currently being built in Ford of Europe are more decision orientated than previous batch simulations. They have incorporated in them interactive routines to input and change the variable data of the model. This is usually done using a paging facility where the screen is completely cleared and the current data values displayed. The cursor is then used to overwrite the old values with the new data and the interaction validates the data as it is entered. Error reports and help facilities are standard features of these interactive routines.

Recently, some simulation software packages have been developed that offer an improved method of displaying the model's status. With these packages the screen acts as a window onto a number of large pictures of information. The user can select which pictures to display on screen and set the size and base coorindates of each picture. This provides a more flexible method of displaying information than using paging. Furthermore, recent advances in the performance and functionality of the graphics capabilities of microcomputers should result in further improvements in visual interactive modelling packages.

As previously noted, a most important feature of the simulation models developed in Ford of Europe are the interactions that allow the users to change the model variables. Interactive routines are usually incorporated to allow preprogrammed 'what if' options to be taken up and assessed at any point in the simulation run. In addition, the interactive nature of the simulation allows control to be returned to the user at any critical point in order to highlight a problem that has arisen but has not been catered for in the existing logic of the model. Even at this point standard software routines may possibly be used to effect a decision that would have to be taken in reality and the simulation can be allowed to continue.

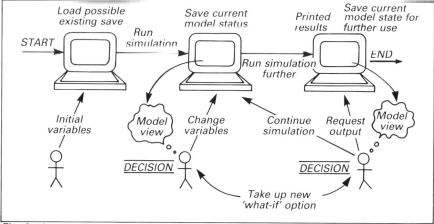

Fig. 3   Interaction sequence

In general the interactive sequence of user decisions may be represented as in Fig. 3. The incorporation of more and more alternatives at decision points in a simulation inevitably increases the amount of checking and shuffling of information that occurs as the model is running. This slows down executive time. A number of run-time improvement techniques have been used successfully in large models written by Operational Research. These include Fibonacci searches of sequential stores and binary heap sorting methods. Also the conversion of real numbers into integer equivalents before storing and sorting and then reconversion to real only on output has been found worthwhile. These techniques are transparent to the user beyond their associated improved run-times.

A further effect on the final simulation model design from the use of visual interactive techniques has been the virtual replacement of a final report on the area under study with a user guide produced by the model builder. The user guide contains complete instructions on how to use the model, what facilities are included, what assumptions have been made and what interactions are available. It may also include some guidelines for experiments to be performed but usually it contains no detailed recommendations of operating strategies for the area being studied. This is now done entirely by the trained industrial engineer in the user area.

Other benefits of visual interactive models

The use of visual interactive modelling software has lead to several unforeseen benefits and this section briefly describes some of them.

Personalising the model

Standard interactions are available to change the display position and colour of items that appear on the screen. The users can modify the screen to their own personal taste and this allows the users to make a contribution to the design of the model. Consequently, the users feel that the model belongs to them, thus generating greater enthusiasm and commitment to the model.

Models as catalysts

Visual interactive models quickly showed that they could act as a powerful stimulus to the user's imagination. Furthermore, groups of users with conflicting interests have used the models to understand each other's views and help resolve their differences. Thus a visual interactive model promotes new and original ideas and can act as a catalyst for discussion between groups.

Models as training aids

Visual models are used to introduce line foremen and supervisors to new facilities and working methods. The model provides a cheap and safe method of giving foremen experience of running their plant. It is the industrial equivalent of a flight simulator.

Productivity

As the use of visual interactive simulation modelling has increased within Ford of Europe, demand for the development of further models of the company's manufacturing facilities has increased. With more people than ever developing models it could reasonably have been assumed that the overall demand would fall; this has not been the case and the backlog is increasing.

To help alleviate the backlog, the department has put effort into the development of user-controlled generic models. To date two such models have been developed and their impact has been impressive. With one such model the time to construct a 30 station machining transfer line model has reduced from approximately two and a half months to five days. A case study based on the machine transfer line generic model is described later.

The need to cut down the amount of time spent programming has also led to generic model construction which is data driven directly from the keyboard. This is now being supplemented by the use of structured data files which are read into the model during the definition and execution phases. Both these approaches feature dynamic screen layout redesign during normal running.

Another approach adopted has been to review the simulation software market place on a regular basis. This ensures that any products which do become available can be evaluated quickly and all current simulation software suppliers to Ford of Europe kept aware of how they should be developing their product. In addition, should alternative simulation software become available, which offers significant productivity gains whilst retaining interaction and screen design flexibility, then a change of host simulation software would become inevitable.

Case studies

This section gives two case studies that illustrate the range of models developed within Ford of Europe. The first study is an example of a model dedicated to improving the capacity of an existing facility. It is typical of the projects undertaken to model both planned and current facilities. The second case study briefly describes a generic model that allows users to build their own models without the need to program.

V6 crankshaft machining line – Cologne

The V6 engine is installed in cars built in Europe and light trucks built in North America. The components for the engines are machined and the engines assembled in Cologne, West Germany. The demand for V6 engines is high and on occasions the demand exceeded the rate of manufacture. Studies of the production facilities had shown that the crankshaft machining line was the major bottleneck in the supply of parts to the V6 engine assembly lines. To improve the supply of crankshafts a major investment in automation for the V6 machining line was made. The increase in output anticipated from the investment did not fully materialise and a visual interactive simulation of the current line including automation was requested so that methods to improve capacity could be investigated.

Normally the effect of introducing automation to parts of an existing production line would be understood before installation through the use of a simulation model. In this case, however, the availability of visual interactive simulation was not known until after the automation was introduced. The model simulated the current manufacturing process and allowed for new crankshafts known to be needed in the near future. The automation between machines was modelled in detail to investigate its full effect and alternative strategies were available through interactions so that users could carry out experiments.

Investigation of the existing machine line revealed that over 50 machines were used on the line along with a number of dedicated mechanical loaders. Also seven automatic loaders were serving more than one machine. These automatic loaders have the ability to distinguish between machine types. The transfer between machines was automated for nearly the complete length of the line with some sections using platens or another type of carrier along the conveyor. The flow of crankshafts down the line was not linear but included a number of junctions where the route taken by the part depended on derivative type and prevailing store sizes at that point. The model included on and off loading of parts between machines, machine breakdowns, and tool changes. Some of the code relating to these aspects of the model was taken from models already developed by the Operational Research Department. This particular model concentrated on the logic prevailing at junction points and the shared automatic loaders to ensure the model accurately reflected the existing line.

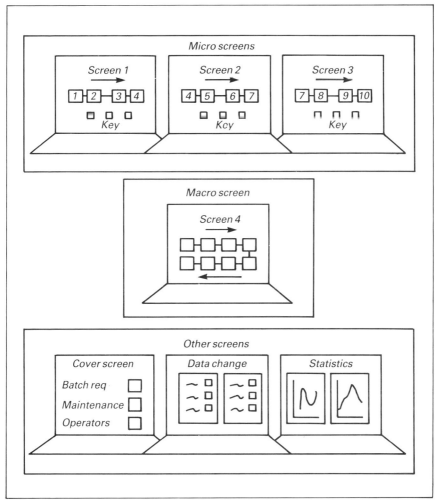

Fig. 4   Screens available for V6 crankshaft line model

The detail of the model required the use of more than one page to display the model. Four screen pages were used, three at the micro level, on overlapping sections of the line, and one macro screen that did not show off-line stores or standby operations but gave an overview of the condition of the line. A full page was dedicated to machine data changes and statistics. An initialisation page was programmed that allowed the user to set up batching requirements and the availability of manual loaders and maintenance men. Fig. 4 shows an outline of the screens the full model offered.

Generic transfer line model

A generic model is a flexible simulation model aimed at a specific family of manufacturing facilities. Extensive interactions allow the user to build significantly different models without the need to program.

Generic models are used by trained engineers to construct models using the building blocks available in the appropriate generic model. Thus one generic model gives rise to many sub models. A generic model of machine transfer lines has been developed and is known as MENTOR.

Machine transfer lines are a common class of facilities found in the automotive industry. They are production lines where parts flow from one operation to the next on conveyor systems that also act as buffer stock between machines. This conveyor automation is called on-line storage. Under special conditions floor space adjacent to machines can also be used to store parts and this is known as off-line storage. The capacity of these production lines is difficult to calculate accurately because of the complex interferences of the planned and unplanned stoppages of the machines that make up the line. MENTOR is used to plan these facilities and predict their capacity.

A structured method for using generic models was developed and users encouraged to follow this process. The first step in the MENTOR modelling process is to draw a flow diagram of the facility and collect the data that is needed to run the model. Once the specification is completed the model can be defined and checked out to make sure that it is correct. The final stage is to experiment with the model in the usual way.

Many people contribute to the specification of a MENTOR model. For example, information from process, layout, and industrial engineering is required to build a typical model. It is most important that their contributions are accurate, acceptable, and completed on time. In order to satisfy these criteria it was necessary to develop a method for collecting information that was clear, consistent, and ensured that the model was built quickly. This was achieved by using a parts flow diagram to represent the physical and logical relationships of the machines and conveyors that made up the facility. The flow diagram uses standard notation to represent on paper the building blocks (machines, conveyors, stores, etc.) and logic elements (joins, splits, assembly points, etc.) that are available within MENTOR. Reference names are given to each building block and an appropriate data sheet completed. The data sheet supplements the parts flow diagram with information about the physical characteristics of each building block (cycle times, store sizes, etc.). The data sheet also details the rules which govern how this building block is fed by preceding blocks and how this block feeds following blocks.

It is beyond the scope of this paper to describe fully the parts flow diagram and data sheets; their importance, however, should not be underestimated. They form the most up-to-date description of the problem against which all contributors discuss their experience, provide data, and confirm that information is correct. A futher benefit is that drawing a parts flow diagram guarantees that a model can be built. This is because the diagram's notation only contains those building blocks and logic elements that are available in MENTOR.

The next stage in the process is to define the model using interactions that specify the logic of the transfer line, the picture on the screen, and the model data. All interactions are controlled from menus that are written in a clear and consistent way.

Particular attention was paid to the screen definition phase because this interaction had to be flexible enough to represent a complex layout but still easy to use. The balance between these two aims was achieved by presenting output from the model in different ways. Standard shapes representing the different types of building blocks and the flow of parts through the model were developed as a method of displaying the dynamic screen. These shapes can be placed anywhere on the screen page and they change colour during the model's progress showing the changing status of the building blocks. In addition, the user can enhance the display by using text and graphics supplied from a data file.

Other utilities with MENTOR allow the user to collect production and utilisation statistics, and display them on the screen as histograms, time series or tables. When the user has finished the model the status of the simulation is saved on to the microcomputer's hard disk.

The final stage of a MENTOR study is to run the MENTOR execute model. This is the part of MENTOR that contains the code for the events that simulate the action of the building blocks in the generic model. The starting point for carrying out model run is to restore the saved status that was set up in the define stage. A series of controlled experiments can then be carried out. The user has the ability to change all of the model data at any time during a model run and can also request printed reports of the model statistics. MENTOR models have been used to examine ways of making more effective use of labour, reducing stock levels, and quantifying the effects of different tool-change strategies.

There have been many benefits from developing the MENTOR modelling system. The most important of these is that visual interactive models of transfer lines can now be built within days rather than months. Inexperienced users can develop models without the need to learn a programming language, and code from MENTOR has been incorporated into other more complex models.

Future plans for improving generic models are in two main areas. Firstly, to reduce further the time spent on data input and the presentation of results efforts will be made to integrate these models with standard database, word processing, and analysis packages. Secondly, ways to improve the model experimentation process will be examined.

Concluding remarks

The work described represents only a small part of the total effort expended in this area over the last four years. All involved in the work are enthusiastic and well motivated.

Visual modelling has opened up a new dimension in the communication process not only between model developers and users but between users and their own management. This communication process has also extended downwards to such an extent that line workers are being given the opportunity to view and interact with the models.

Visual interactive modelling is still in its infancy. Over the next few years there is likely to be an accelerating development; during this period Ford of Europe will maintain its leading position.

References

[1] Hurrion, R.D and Withers, S.J 1982. The interactive development of visual simulation models. *J. Opl Res. Soc.*, 33(11): 973-975.
[2] Fiddy, E., Bright, J.G. and Hurrion, R.D. 1981. SEE-WHY: Interactive simulation on the screen. In, *Proc. of the Institute of Mechanical Engineers*, C293/81, pp.167-172.
[3] Hollocks, B.W. 1983. Simulation and the micro. *J. Opl Res. Soc.*, 34(4): 331-343.
[4] Bell, P.C. 1986. *Visual Interactive Modelling in 1986*. Working Paper Series No. 86-08. School of Business Administration, University of Western Ontario, London, Canada.

COMPUTER SIMULATION FOR FMS

N.R. Greenwood
General Electric Industrial Automation Europe, UK
P. Rao
General Electric Corporate Research and Development, USA
and
M. Wisnom
Structural Dynamics Research Corp., UK

Optimisation of system design is a key problem in planning flexible manufacturing systems. This paper demonstrates the viability of sophisticated factory simulation packages within a programming-free environment. The system developed by General Electric is taken as an example of meeting the ever-increasing requirements of advanced systems designers.

Automation no longer means the simple like-for-like plant replacement that has been traditional for many years throughout most manufacturing industries worldwide. Now, by automating, a company frequently commits a substantial quantity of resources to a facility, the complexity and likes of which the company has probably never seen before. Immense risks of almost every conceivable type are associated with these strategic, long-term investment decisions. But despite all this, the popularity of flexible manufacturing systems (FMS) continues to grow in leaps and bounds. This is particularly noticeable in Europe where the harsh commercial pressures of survival have led to the development of FMS technology which in many ways is more advanced than anywhere else in the world. As a result, manufacturing technology is forced to continue to progress and, theoretically at least, the engineering risks become fewer and fewer.

Nevertheless, three substantial engineering hurdles remain for the potential FMS designer:

- How does one produce an integrated system from the varied equipment frequently present?
- How does one avoid the very high once-off software investment?
- How does one 'optimise' the design of the proposed system?

Most manufacturers of FMS equipment, or substantial users of FMS (and General Electric in America is very much in both these categories) are investing significantly to address these issues. While considerable progress has been made in all these areas, it is the third category which is of particular interest to this paper. Namely, how does one optimise FMS design? Specifically, what contribution can computer simulation make to this process?

Use of simulation

Simulation of discrete parts manufacture and continuous processes has been practised in a relatively sophisticated manner for over two decades. Predicting system bottlenecks, utilisation of fixed and movable resources, work-in-process and production rates – this type of modelling of manufacturing system performance usually involves the use of a high-level 'language'. The model itself has to be programmed using the particular syntax relevant to that language. This usually turns out to be a laborious and time-consuming process at best, frequently resulting in models which can only be understood by the programmer, and reams of results which are difficult if not impossible to interpret. More recently these languages have improved in their functionality with the addition of constructs (i.e. special subroutines) to handle special manufacturing situations such as materials handling. However, they still suffer from a lack of user-friendliness.

Recognising the current trends in graphics and animation, and the concurrent advances in computer science, simulation software developers have started to interface real-time graphics and animation to the traditional simulation languages. Graphic pre- and post-processors to simulation languages have become available over the past few years, greatly improving the interpretation of simulation output. In fact, most people would probably now agree that a few minutes of graphic animation of critical output sequences can eliminate the need for scanning several hundreds of pages of computer printout (which typically are filled with rows and rows of numbers indicating time-dependent machine performance, queue content and transporter location). Relatively sophisticated output analysis packages involving both graphics and animation are now available. However, the general trend is still towards using a simulation 'expert' to create the model, thus allowing the person or group ordering the simulation to become merely an observer of the output as defined, and provided by, the simulation expert.

In this age of rapid information transfer, it is not difficult to understand that the simulation user desires not only to define the problem but also to input design parameters in a convenient manner, execute the simulation, analyse the results, and then maybe change the model and try again. All this with the very minimum of effort.

Within General Electric it is felt that the era of programming exclusively by an expert is coming to an end. The microcomputer revolution is placing incredible computing power in the hands of individuals who, in the past, have had to depend on programmers to build computer models for them. Now the ultimate users of simulation results can be given the ability to build and execute models by themselves. All it needs is an appropriately smart, programming-free environment.

This is the goal that General Electric's Corporate Research and Development Center in Schenectady, New York, has had in mind for the past few years, with its factory simulation program. Out of this extremely well-equipped facility, which has available virtually every recognised significant simulation tool, has come a number of stand-alone application modules which provide a highly sophisticated programming-free modelling environment for the user.

The application modules handle four specific types of manufacturing situation. Namely, serial processing, assembly loops, the job shop and materials handling. These basic modules are characterised by an entirely programming-free environment involving very user-friendly graphics and menu-driven input. All of this results in a simulation which runs totally transparent to the user, and which is followed by different levels of graphical output.

From these four basic modules it has been possible to develop two 'super modules'. One combines serial processing and assembly loop operations for handling assembly facilities. The other combines the job shop and materials handling modules, to model machining-orientated flexible manufacturing systems.

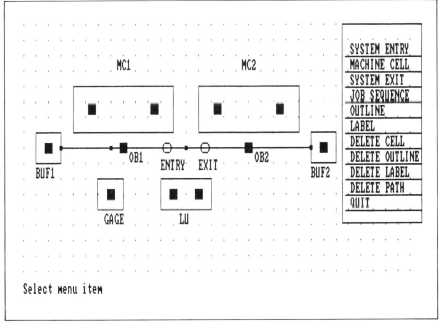

Fig. 1   Example FMS layout

Example of FMS simulation

To give an indication of how powerful the above-mentioned tools are (which, incidentally, is virtually impossible without the aid of a colour graphics terminal), an example of the simulation of an imaginary (though seemingly popular configuration) flexible manufacturing system will be described.

A scaled, simplified layout of the facility is produced as part of the user input with either a 'mouse', or the keyboard, as shown in Fig. 1. This system comprises four machining centres, grouped together in cells of two (i.e. MC1 and MC2). These are served by a single rail-guided transporter, whose true path may be defined by pointing the 'mouse' at the various workstations, and 'clicking'; followed by drawing the defined path using 'rubber banding' graphics. The other workstations include a load/unload area (LU) consisting of two work areas, a single coordinate measuring machine (GAGE), two on-line buffer storage areas (BUF1 and BUF2), and two output buffers (OB1 and OB2) for machine 1 and machine 2 respectively. The constraints on queue size were defined as follows:

MC1	=	2	MC2	=	2
OB1	=	2	OB2	=	2
BUF1	=	2	BUF2	=	2
GAGE	=	1	LU	=	unlimited

The queues were preloaded to half their capacity at the start of the simulation run.

The system was simulated to run for three eight-hour shifts during each day, for five days a week. But the load/unload station was manned only for the first two shifts of each day. In addition, it was assumed that all the machines would require, on average, one hour of preventative maintenance during each week; and that this would be carried out on a random basis.

The cell was to process three types of jobs, with a total demand of 80 of each per week. Material to be processed was assumed to arrive at regular intervals throughout the three shifts. The processing sequence for each of the three jobs was as follows:

    JOB1: LU, MC1 or MC2, GAGE, MC1 or MC2, GAGE, LU
    JOB2: LU, MC1 or MC2, LU, MC1 or MC2, GAGE, LU
    JOB3: LU, MC1 or MC2, GAGE, MC1 or MC2, LU, GAGE,
          LU

It was assumed that loading and unloading (or refixturing) would take 15 minutes, that the machine cycle times would vary from 5 to 20 minutes depending on the operation, and that the gauging of the parts would take 15 minutes. Also, it was assumed that the length of cell was 100 feet, and that the single transporter (with a capacity of one part) moved at a speed of 100ft/min.

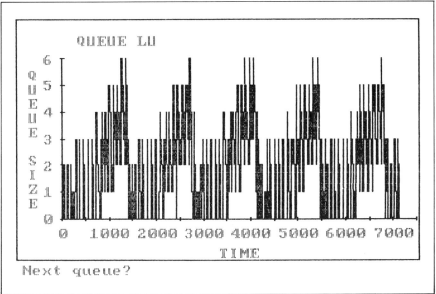

Fig. 2   Load/unload station queue vs time

As mentioned above, simulation input consists of a simplified graphic layout drawn on the computer VDU by using a 'mouse' or the keyboard. This is followed by a menu-driven question-and-answer session for the input of information such as machine cycle times, maintenance, breakdowns, etc., based on the make-up of the system as defined in the layout.

The simulation then runs totally transparent to the user, not having required the generation of any computer code for the execution. The simulation is automatically configured, based on user input. Results of the simulation may be examined in the form of a summary report, or as plots of queue content against time. For this discussion, the queue contents of one week of operation are shown as Figs. 2–4.

Fig. 2 shows the queue in front of the load/unload station (LU). It is interesting to note the periodic increase in the input queue caused by material continuing to arrive during the third shift, when no operator is present to load the work into the system. The two on-line buffer stores (BUF1 and BUF2) are rarely utilised, except at the beginning of each day, when machine utilisations are necessarily 100% (caused by the overloading of the input queue; see Fig. 3). This phenomenon may be observed clearly in Fig. 4, where the single-capacity queue (GAGE) in front of the gauging maching is continuously full at the beginning of every day (indicated by dark vertical bands), and only partly full at other times.

While this example has been rather oversimplified to help demonstrate the ease of use of the simulation package, rather than the feasibility of this particular FMS design, it is nevertheless easy to see how the use of the model very quickly indicates ways in which the

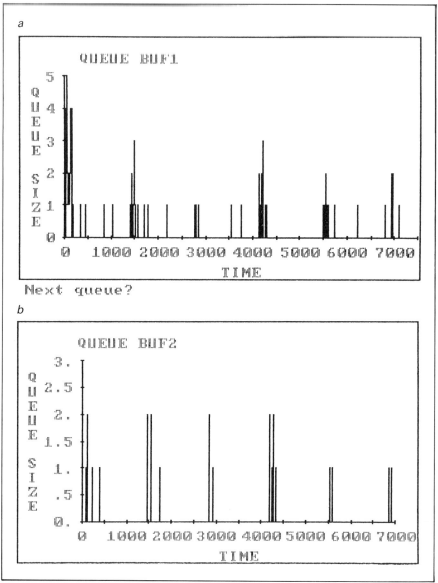

Fig. 3   *On-line buffer store contents vs time: (a) store 1, and (b) store 2*

overall efficiency of the system may be improved. Certainly, a major increase in user productivity may be achieved by using the GE simulation package, especially once the relevant data have been collected. Although a normal simulation model of such an FMS installation using a more conventional general-purpose simulation language could take several days or even weeks to write (even if graphics are available), the time being very dependent on the skill of the operator, the model described above took less than an hour to input, run and interpret.

Fig. 4   Measuring machine (GAGE) queue contents vs time

Concluding remarks

The main purpose of this paper has been to demonstrate the viability of sophisticated factory simulation packages within a programming-free environment. In particular, the system that has been developed by General Electric, which is believed to meet admirably the ever-increasing requirements of advanced automation systems designers, has been significantly improved in its functionality (indeed the authors are presently working on an updated version of this paper). A number of micro-based packages are already on the market, but it is hoped that the reader will appreciate that the GE system offers substantial advantages over its now ageing predecessors.

The GE packages were designed primarily to run on the DEC Vax range of computers, but it has since proved possible to implement a significant proportion of the software on the IBM personal computer and on appropriately compatible machines such as the Data General ONE.

It is GE Industrial Automation's intention to make these packages available in Europe in the very near future.

FMS: WHAT HAPPENS WHEN YOU DON'T SIMULATE

J.E. Lenz
CMS Research Inc., USA

One of the most discouraging aspects of flexible manufacturing systems (FMS) is the huge difference between design performance and actual performance. This might be expected for the first few such systems, but after more than ten years of technological improvements the gap seems to be widening instead of closing. The primary reason for this can be placed upon the design tools which have been used. The weaknesses of inappropriate design tools are highlighted and it is suggested that computer simulation must be an integral part of any FMS design. First, the traditional design tools for automation are discussed, followed by their application to FMS design. Many examples of such practice are presented, together with some of their disastrous results. Finally, a proper FMS design procedure is proposed.

Traditional techniques used for the design of transfer lines, automated storage/retrieval systems, and robotic work cells have often been applied to the design of flexible manufacturing systems (FMS). These techniques, which range from balancing the line to fixed equipment cycle times, evaluate from a steady-state viewpoint. This view is adequate for synchronous equipment since the output of one component becomes the input of another. However, this simple relationship between components is not consistent with the philosophy of FMS. In fact, when these techniques have been applied to FMS design, disastrous results have occurred. These results range from building more than twice the number of pallets needed for the system to proposing a seven-cart material handling system when one cart could do the job.

The reason for these mistakes is the design technique's inability to study the interaction between system components. When outputs of one component are inputs to another, the interaction between components is well defined. However, when this characteristic does not apply, as is the case in FMS, use of such design tools only provides a view at 'optimal' steady-state conditions.

These 'idealistic' conditions provide analysis of the capacity of the system, but 'overstate' its realistic production capabilities. This 'overselling' of FMS has cost vendors and customers millions of dollars and has caused it to fall into disfavour with some industries. It is time for FMS design to focus on realistic performance, and this can only be accomplished by a detailed study of the system's operation.

On account of the 'flexibility' of many FMS components interacting with many others, the simple input-output relationship cannot be defined mathematically. Thus a replication or simulation of the system's operation is necessary to study inter-component relationships. Through such study, FMS designers will learn of the strengths and weaknesses of systems, and users will be able to devise realistic production plans.

The majority of the material for this paper comes from five years of consulting in FMS design and operation. All examples are real, but names and other identifying information have been altered or avoided. Such information is subject to confidentiality, so references for the examples are not given.

Traditional design for automation

Factory automation prior to FMS has included transfer or assembly lines and automated material handling systems. The design procedure is highlighted for each of these areas in the following sections.

Transfer line design

Many FMS designs are done through procedures consistent with the design of a transfer line. This can be attributed to two primary facts. First, most of the vendors who design FMS have been involved in factory automation from its inception, and most of this automation was transfer lines. Thus designers have been schooled into thinking of transfer line operation. Secondly, designers with strong backgrounds in transfer line automation tend to bring their known design tools into the FMS design.

In the design of transfer lines two general strategies are followed: balance the line or line-of-balance. In either case the objective is to define each workstation's 'cycle' to be as close as possible to all others. That is, a perfectly designed transfer line will have all stations with identical cycles and zero slack per station. With this objective, 100% station utilisation is achieved. This is realistic since all inter-component relationships are deterministic and unforeseen 'surges' are not possible.

Tools which assist in such a design include machine loading calculations and the number of pallets. The number of pallets can easily be identified as equal to the number of 'positions' in the transfer line. These positions might only be stations but could include some intermediate stops as well.

Calculation of the machine loadings can be done by using

mathematical expressions including production volumes and cycle time. Since the cycle times are nearly identical for all stations, the loading is approximately the same for all stations.

With these two results, a transfer line can be designed and proposed. There is no need to worry about control or congestion developing during operation since each part takes the same path through the system.

When this technique is applied to FMS design, the planned system will have enough pallets to maintain 100% station utilisation because this is the underlying characteristic of a transfer line. In reality, however, no system will maintain 100% station utilisation, and as a result too many pallets will be included in the system. This only becomes a problem during operation since, in the case of a transfer line, 100% station utilisation is never achieved either. However, 100% station utilisation is achieved while the line is running. Also, when one station fails, the entire line goes down. In FMS operation though, one station failure does not stop the entire line, but when all stations are designed for 100% utilisation a queue (or congestion) will develop at the down station.

If FMS are designed with transfer line characteristics, then they should be operated with transfer line characteristics. The following are examples of FMS which were designed with transfer line design tools.

Determining the number of pallets for an FMS. A system was designed for a heavy equipment manufacturer by using the machine load calculation used in transfer line design. When the number of machines was determined, a layout of machines around a bidirectional track was proposed. When it came time to determine the number of pallets for the system, the number of pallet 'positions' was counted. The number of pallets which were proposed and constructed was one less than the number of pallet 'positions' in the system. When the system became operational, it was found that only 60% of the pallets could possibly be entered because of trafficking. In other words, 40% of the pallets built could never be used by the system.

Design for 100% station utilisation. An FMS was designed with a uniform load over all stations. The number of pallets proposed for the system was sufficient to maintain this 100% station utilisation. When the system became operational, its production ranged from 10 to 80% of the proposed figure. These large swings in the production rate were due to single station failures and 'surges' of pallets waiting for down equipment. As a result, the line became 'out of balance' and high inefficiencies of stations occurred. After some period, this 'surge' would work itself out of the system, only to recur with the next failure. The system lost production capacity not only during periods of down equipment but also during this recovery from the 'surge' which always resulted. In this specific case, production could actually be increased if the FMS were run as a transfer line. That is, when one station failed, no pallet motion would be permitted until it was repaired. In this way the lost production would be tied to machine failures and not to the time required to get the line back into balance.

Material handling system design

The design procedure for an automated material handling system involves an estimation of the average transport cycle time. This time includes the time to load and unload the transporter and some moving time determined by the average distance. In many systems, the pick-up and deposit points are the same and often the path is the same for all transporters. Thus these design procedures can be very accurate for computing the load of the transporters. The primary result of this study is to determine the number of transporters needed.

To apply this procedure to FMS, an average cycle for the transporter needs to be derived. However, the paths are not always the same and the 'average' becomes a mathematical, but not necessarily accurate, result. Another method for determining the number of transporters for an FMS is to count the number of 'zones' and use some percentage of this number. For example, a wire-guided layout might permit 20 specific areas where a single cart can reside. The maximum number of vehicles might be ten carts (i.e. 50% occupancy). Usually, the number of vehicles proposed is always near the maximum which the system can handle.

The following are examples in which FMS installations were designed with these material handling system design tools.

The $1,000,000 transportation system. An FMS was proposed for a system with 14 stations. The material handling system was a wire-guided vehicle system with seven carts. This system added $1,000,000 to the total cost of the 14 machines, which was $2,500,000. An alternative vehicle system was evaluated which used a single cart moving bidirectionally. Computer simulation was used to arrive at the design characteristics of the vehicle and a request for quotation was initiated. The cost for the single vehicle with a maximum speed of 450ft/min and 10s pallet transfer was less than $400,000. This saving of more than $600,000 did not include the fact that now only 50% of the number of pallets were needed because of the reduction in transportation time. This saving of 50% less work-in-process will pay each day the system is in operation.

No room left to move. A transportation system was designed and implemented into an FMS which included the maximum possible number of vehicles. This maximum number was determined from design techniques used for traditional material handling system design. However, when the system became operational the dynamic paths used by the vehicles resulted in 'deadlocks'. That is, carts would bunch up in certain areas of the system. In some cases ten vehicles would be waiting for a single vehicle to move. This traffic congestion lengthened the average transport cycle. One solution was to add more transporters. But when the real problem was understood, removal of several carts relieved much of the congestion and the transport cycle decreased, resulting in an actual increase in transport system capacity.

FMS: The dynamic areas

The number of stations and transporters can usually be determined from steady-state analysis. But even these are subject to error in some unique situations. However, there are many additional areas which need evaluation, and all of these deal with the dynamic nature of FMS operation.

Simplified routing

As the 'flexibility' of transporting a part to any one of several identically tooled machines is a common characteristic within FMS, there is a need to study the ramifications of this 'flexibility'. One ramification is the number of different paths a part can take through the system. For example, suppose ten pallets are available for a part which must be machined at one of three machines, then one of four machines. The number of different paths for any part is $10 \times 3 \times 4$, or 120. The FMS is expected to produce the same quality no matter which combination is used. That means that 120 different paths must be tested and adjusted to ensure uniform quality.

When these high numbers exist, it is necessary to simplify this operation to a manageable level. One way is to limit the number of pallets which can accept a certain part type. For example, instead of ten pallets suppose only two are permitted. Now the number of paths is 24. Before this can be accepted, the impact of such a restriction must be identified as to the effect upon machine utilisation and production rates. It might also require running more part numbers simultaneously, and thus a greater tooling capacity for the machines. An application of this technique follows.

A large FMS was proposed with 40 pallets and 18 machines. The number of different paths for any part was over 2,800 through the system. A method to reduce this number was essential to maintain part quality because it would be impossible to guarantee that all paths would produce identical parts. Computer simulation was used to simulate restrictions of certain pallets to specific machines and certain parts to specific pallets. The result was a reduction from 2,800 paths to less than 100. Even the 100 is a high number, but any further restrictions resulted in improper production ratios and inefficient use of stations. This system is now being installed with these 'logical' restrictions.

Another system was designed using a conveyor-type material handling system. From its layout and design, it looked like a good solution to a complex material handling problem. However, when a computer simulation was run, it was found that more than 50,000 motions were required in the conveyor each eight-hour shift. This is equivalent to saying that more than 100 limit switches needed to be made during each minute. Once this characteristic was known, little confidence remained as to the likelihood that the conveyor would work without a reliability problem.

Computer control compatibility

A critical area for FMS design is the compatibility between control strategies and hardware configurations. When FMS managers are consulted, they will commonly refer to the control as the major weakness of the system. The reason for this is that the control is not included as part of the system design and is assumed to be adjusted to meet the needs of the system. 'It's only software, and we all know how easy it is to change software'. Indeed it is easier to change software than to replace a machine when it doesn't work, but in many cases the sophistication of the control is severely limited by the system configuration.

For example, an FMS was designed and installed using two vehicles on a single bidirectional track. These vehicles could not pass one another but each was needed to serve all stations since parts could be routed from one end to another. It was assumed that someone could write an algorithm which would 'optimise' the movement of the vehicles to avoid interference. However, the algorithm which resulted was for steady-state operation and used the ability to predict future activities. The dynamics of the system never made steady state a reality and thus the cart control algorithm was useless. To resolve this problem, only one vehicle was put under computer control at a time, with the other remaining under manual control.

A design procedure for FMS

The primary objective of FMS design is to configure a set of machine tools and auxiliary equipment which has 'realistic' capacity to meet a planned production at the lowest cost. The portion of this objective which makes it challenging is the word 'realistic'. Here such occurrences as congestion, machine breakdowns, operator intervention, part shortages, variable batch sizes and tool replacement must be allowed for to ensure an accurate relationship between the design and the installation. The procedure described below focuses on this requirement.

There are three steps in FMS design: aggregation, simulation and animation. These steps are intended to be hierarchical, and as the design moves through these three steps more is learned about the system and fewer choices become available. The first step is to 'ball park' the system. With each succeeding step, further detail of the system configuration is provided. Also, each step must have a 'realistic' means of measurement.

Aggregation

This is the first step in FMS design and its primary objective is to provide a 'feasible' system configuration. A feasible system includes sufficient numbers of machines, transporters and pallets with aggregate capacity to meet the production requirement. The planning horizon for this step is usually no shorter than a month and might be as long as a year.

Design tools for this step usually include some mathematical model for FMS from which data can be provided and individual component utilisation can be computed. Two such tools are CAN-Q[1] and SPAR[2]. With both of these tools, the data input includes a production requirement for each part type and a process plan (operation list with estimated cycle times). In addition, an approximate number of stations and transporters is entered. It is not essential that the correct number of stations or transporters be entered initially since these tools provide direct output indicating how many components are required.

In this step of FMS design, gross requirements of production are compared with gross capacities. This comparison is often referred to as aggregate planning. However, for proper application to FMS, 'realistic' margins for downtime and system inefficiencies must be allowed for in the use of the tool. First, the aggregate station requirements are computed. These are done for each station group (tool alike machines) by adding operation times for the frequency of production over all part types. This requirement is then compared with the gross capacity of the station group, which is computed as the planning horizon times the number of stations in the group. It is at this point that many designers wish to use the transfer line approach to the FMS design problem.

In a transfer line approach, downtime allowance is chopped off the planning horizon. It is natural to do this since when one station is down the entire line is down; and when the system is operating, all stations are operating at maximum utilisation. This characteristic is emulated when the planning horizon is reduced to the 'breakdown allowance' factor. However, this is not characteristic of an FMS. When one station fails, the entire line does not stop but continues to operate. Therefore reducing the planning horizon by, say, 15% does not result in a true representation of the FMS and how it will eventually operate. The proper representation is to study the entire planning horizon but to restrict station utilisation to some specified level. Each station group should not be loaded over the "realistic" time which these stations can be expected to be operational during the total production period. Also, this level need not be applied uniformly to all station groups. This procedure will provide an accurate picture of how long each station group is required to be operational in order to meet the planned production capacity.

Once the number of stations which does not overload beyond some level has been determined, the load of the transport system can be computed.

The aggregate requirement of the transport system is computed by counting the number of 'moves' a part requires multiplied by the production requirement and an 'average' transport cycle. This requirement is compared with the capacity, which is computed by multiplying the number of transporters by the planning horizon. When the capacity is compared with the requirement, a 'realistic' percentage of utilisation should not be exceeded.

The final area of aggregate planning for FMS is evaluation of the in-process storage capacity and determination of the number of pallets. To determine the appropriate number of pallets, a pallet cycle must be estimated for each part type. This cycle includes total operation time, transportation time and in-process storage time. Once this cycle has been estimated, the number of parts which can be produced by a single pallet is computed, which is then divided into the production requirement to determine the number of pallets for each part type. The total number of pallets can be compared with the in-process storage capacity of the system.

Once the aggregate planning has been done for an FMS, and 'realistic' measures have been incorporated in the capacity, the FMS design will contain the correct number of stations, transporters and in-process storage positions. Also, a good estimate for the number of pallets will exist. Once these numbers have been accepted, the next step in FMS design is to identify how much congestion results from integrating all of these components together. This study of congestion is done in the second step of the design process.

Simulation

The study of FMS congestion which might result during operation can only be done by use of computer simulation. Mathematical models are not adequate to study the dynamic nature of FMS operation and to 'foresee' the thousands of situations which will result from normal FMS operation.

Computer simulation of FMS should have as its primary objective determination of the inter-component relationship for congestion. Simulations should be performed for a system which has already been 'ball parked' since this tool is complicated to use and, if the number of machines, transporters and pallets has not been determined, can require many iterations before an operational system is reached. This complexity results because poor system performance might be due to too many pallets, traffic congestion, or poor tooling assignments to the stations. When all of these areas are subject to change, the direction for the simulation study is too varied and solutions to problems often lead to more problems and the iteration loop of simulation becomes a vicious circle.

Simulation will be most effective when the FMS has been properly sized. It must simulate every pallet motion which will take place in the FMS operation. This includes individual motions in the transportation system. Many simulation languages can simulate station and storage activity but only an 'average delay' for a pallet motion in the transportation system. When the detail of individual transporter movement is not included in the simulation, no study is made of dynamic paths, interferences of vehicles, and algorithms for traffic control.

Many simulation languages are available for manufacturing system evaluation, but only a few include the ability to simulate the detailed motion in the transportation system. If this detail is not simulated, then the computer simulation results will not yield much more information than what was found in the aggregation step of FMS design. Just because it is called simulation does not mean that it is solving step-two design problems of FMS.

Animation

The third step in FMS design is to study the compatibility between the FMS hardware and computer control. Animation provides a visualisation of the FMS operation long before it is acutally installed. In this visualisation of the operation, all real-time decision-making can be studied for 'efficiency' of component utilisation. Such algorithms as part balance, station selection and traffic control can be seen in operation and the designer can evaluate each algorithm for 'optimality'.

An FMS animation requires use of a computer simulation which can produce a record of each activity as it occurs in the simulation. These records or 'events' are then used to drive motion through a background of the FMS layout. The motion seen in the animation is caused by 'blinking' the pallet and transporter images through various positions in the background. The order of the blinking is determined by the order of events in the simulation. The animation can be synchronised with the computer simulation or can be done via 'playback' means. The playback is accomplished by recording the event data into a file which is used once the simulation has stopped.

Concluding remarks

The design procedures for FMS have been derived from procedures used for previous forms of automation. These include procedures for transfer line design and automated material handling system design. But when applied to FMS design these procedures fall short in the ability to evaluate the dynamic nature of FMS. The inter-component relationship is not restricted by design and so the design procedure must evaluate the operation of the FMS. A three-step approach to FMS design has been proposed here.

The three steps of FMS design are aggregation, simulation and animation. Aggregation is to identify aggregate requirements and compare these with aggregate capacity. Three areas for comparison are model stations, transporters and storage. The objective of this step is to produce a feasible design.

The second step of FMS design is computer simulation. The objective of this step is to replicate the operation of the system for the purpose of identifying where congestion will develop. If a feasible system is not used initially, the task of overcoming congestion while the number of transporters and pallets is being determined is complex and causes chances for errors in the design process. The simulation must also replicate the real-time algorithms which will be used in making control decisions.

These algorithms can be evaluated in the third step of FMS design, which is animation. This is the visualisation of the computer simulation with all decisions being shown graphically. The purpose of this step is to evaluate the 'efficiency' of various algorithms for the system configuration.

With this three-step approach, FMS can be evaluated correctly prior to installation, and when 'realistic' measures are applied the design performance can become reality.

References

[1] Solberg, J. 1981. Loading and Control Problem for an FMS. Purdue University, West Lafayette, IN, USA.
[2] Lenz, J. 1983. *SPAR User Manual*. CMS Research Inc., WI, USA.
[3] Anon. 1983. Computer graphics – A tool for verifying system design. *Modern Material Handling*, 38(March 7, 1983): 60-63.
[4] Fishman, G. 1978. *Principles of Discrete Event Simulation*. Wiley, New York.
[5] Kaufman, A. and Hanani, M. 1981. Converting a batch simulation to an interactive program with graphics. *Simulation*, 36(April 1981): 125-131.
[6] Lenz, J. 1984. *Using Animated Simulation to Model FMS Designs*, Vol. 1. Auerbach Publishers, Pennsauken, NJ, USA.
[7] Lenz, J. 1985. *MAST and MAST Animation User Manuals*. CMS Research Inc., WI, USA.
[8] Young, R. 1981. Software control strategies for use in implementing flexible manufacturing systems. *Industrial Engineering*, 13: 88-96.

CHOOSING AND USING A SIMULATION SYSTEM

R. Griffin
Inbucon Management Consultants Ltd, UK
and
A.H. Warby
IBM Ltd, UK

Computer simulation is being increasingly accepted by industry as a valuable and practical technique. This paper deals first with the characteristics of simulation and the justification for using it in planning new manufacturing systems. As a basis for comparison the requirements of an ideal simulation facility from the non-specialist user's point of view are discussed. This leads on to a review of features on available systems, particularly graphics, and the ways in which they would be of practical value on a project. Finally consideration is given to the question of justifying the development of an in-house capability as against the use of outside specialist resources.

Computer simulation in some form is increasingly common today. Almost everyone is familiar with arcade games such as Space Invaders in which various forms of mock warfare may be simulated. Another well known example is the training of pilots and astronauts in highly complex flight simulators. The subject this paper covers, however, is simulation in manufacturing, which may not yet be as popular as arcade games or as highly developed as flight simulators, but is now accepted as a vital management tool in the planning and implementation of advanced manufacturing technology (AMT).

Many other examples of simulation can be found. The prime objective common to them all is to minimise the cost and risks involved in complex situations by allowing the mistakes to be made in a 'safe' environment. While a flight simulator serves to minimise pilot error, simulation in manufacturing aims to minimise time wasted and money spent in implementing new technology.

The technique of simulation in most cases requires the building of a dynamic computer model of a situation or system, which can then be studied giving a realistic view of the proposed real world. Alternative versions of the model can be tested in order to discover by trial and error which configuration suits the purpose best.

Computer simulation in manufacturing
Why use it?

Automation has been a field of great progress over the last century, giving us the capability to mass produce many products. To a great extent it has been applied without the help of simulation until quite recently. However, the enormous advance in computer technology and robotics in as short a time span as five years has brought with it a new brand of more flexible automation. This flexible automation is now being applied in areas such as batch manufacture and assembly where investment in automation was previously marginal or simply uneconomic. This recent revolution in the manufacturing industry has brought with it two major problems. First, the systems involved are both highly complex and integrated, which means traditional planning methods are unable to predict and study their highly dynamic nature. Secondly, the increase in application of this technology has created a shortage of skilled engineers in the field of planning for flexible automation. Engineers who previously planned for piecemeal changes would make standard comparisons of floor-to-floor times, labour costs, improvements in work-in-process (WIP), etc. However, they can no longer use their experience and judgement with this new breed of automation. Or at least if they do the likely result is overcapacity in key resources and severe bugs in control systems, resulting in high costs and ultimately the failure of the system to perform to specification.

Computing power has been the key resource for these flexible systems, so it is logical to use this power to help study and solve the problems posed by the dynamic interaction of many factors that is so characteristic of complex integrated systems. Computer simulation is the key tool available to management to do this.

The benefits of and justification for using simulation

Simulation in common with most computer-aided tools can have varying levels of complexity. Just as computer-aided design can mean work carried out with a simple 2D draughting package or equally with a full-colour shaded solid modelling package, so simulation can mean anything from a simple mathematical model taking a matter of hours to build to a hugely complex 'emulation' model requiring as much as two man-years to create. Fig. 1 shows the range of costs attached to the various levels of complexity of simulation. The justification for a particular level of detail could be made in purely financial terms: the savings as a result of the simulation should be equal to or higher than the cost of carrying out the simulation. However, it is not always possible to quantify savings directly, especially when you are uncertain in what form they are going to be until you have completed the

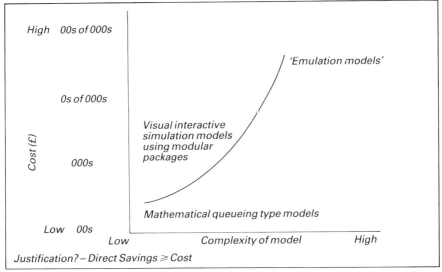

Fig. 1   Relative costs of simulation models

simulation! Experience has shown that the cost of simulations for first-stage planning are rarely greater than 1% of the capital cost of the system. The cost will therefore typically be between £5,000 and £50,000 for investment projects of £0.5 million to £5 million. This of course is not a hard-and-fast rule, but gives a good indication of the magnitude of cost likely to be incurred in getting to a first-stage model that has confirmed the feasibility (or not) of the proposed system. The model will then be suitable for additional detailing in order to test and optimise the system in greater depth. The cost of this stage will depend entirely on the complexity of the system and the level of detail that it is decided is required. In terms of direct savings the cost approximately follows the law of diminishing returns.

The reason for looking at the question of justification so closely is the fact that the costs involved *are* significant and require justifying as the first stage in evaluating any simulation system.

If ultimately it is difficult to quantify the question of 'payback', it should not be forgotten that this now proven and accepted method of planning gives a level of confidence and results that can mean the difference between the system working and not working. Many organisations feel that, despite the law of diminishing returns in terms of direct savings, the high confidence attributable to detailed simulation is the ultimate justification and a small price to pay.

The production engineer's ideal

On two separate occasions recently a senior production engineer explained what the ideal solution to their planning problems would be. In both cases the common ideal was to have the capability of animating a drawing of the proposed system in order to study its performance. At one level it is possible to imagine zooming into a part of a production cell to perhaps study the operation of a particular tool changer, yet at another level all that would be required would be to know the total throughput of the cell per hour.

The implication of these ideals is that the CAD system must have an intelligent knowledge of the behaviour of all the constituent parts of the system and also of how they will behave when linked together. Unfortunately, since this is not the case currently, it leads to the question of what level of detail is strictly necessary to solve the problems these production engineers have in planning their systems and also what *is* available today to help them do this.

What to look for in a system
Graphics
In the examples discussed above, 3D graphical representations were being studied. Indeed there are systems on the market that will animate 3D representations of robot arms, giving the capability of checking for collisions, geometric problems and also off-line programming. However, this type of system is addressing problems of detailed design and not the types of problem concerning complete system behaviour that are the issue of this paper. In the early days of simulation, graphics were not used at all, and the output of the model was purely in tabulated form. Fig. 2 can be used to compare that approach with the approach currently favoured as depicted in Fig. 3. This shows how the intelligent use of 2D block colour graphics can help relate the model to the proposed real world. The use of animation showing material transfer, work flow, etc., further enhances this understanding. We have found that although the use of graphics does not alter the basic simulation logic it gives a level of confidence to both the model builder

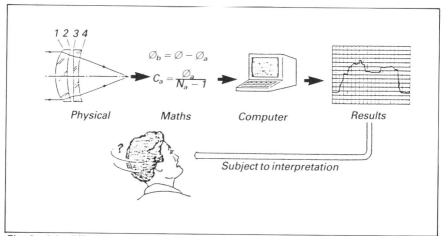

Fig. 2   It is difficult to relate the results of a non-graphical simulation to the system it describes

Fig. 3   Simulation with animated colour graphics – 'seeing is believing'

and other non-specialists that is otherwise impossible to gain. The major advantages of using colour graphics are that:

- It allows the study and understanding of the process and not just the results.
- It gives a simple method of communicating the concepts and operation of complex systems to non-specialists.
- It helps the model builder in debugging system logic as undesirable behaviour can actually be observed on the screen.

It should be noted that some systems that have colour graphics only have the facility to 'play back' any particular simulation run. The ideal situation, however, is to be able to interact with the model as the graphics run. Although animated colour graphics is currently an expensive overhead, the extra dimension it gives in problem solving outweighs this in most cases.

Interaction

The capability of interaction during the model runs is of great importance in the process of both model building and analysis using the model. During model building the system should allow the user to interrogate the status of the model easily in order to compare this with the expected state.

Fig. 4 shows the typical process used in interactively improving model performance and eventually arriving at an optimised system specification. It can be seen that the capability of varying model parameters is a major part of the process, and this should be easy to do using screen interactions provided in the system.

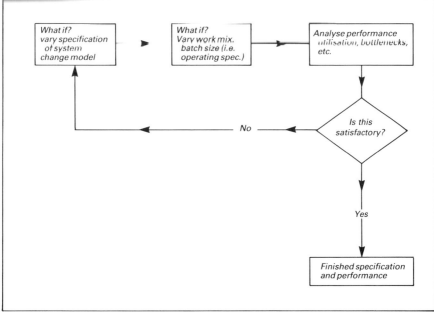

Fig. 4   The process used to analyse a simulation model

The combination of animated colour graphics and slick interactive facilities is the trade-mark of the 'visual interactive system' that currently represents the state of the art in the field of planning for AMT.

A relational database

The use of a database with simulation models is becoming increasingly important as the models being developed are getting nearer to being used as on-line planning tools where large amounts of data are involved.

The database should give the facility to store data flexibly which can then be changed and interrogated easily and quickly. This gives the user the facility to try various strategies of work loading and system specification without worrying about the core logic of the simulation model.

Eventually the use of simulation models as dynamic schedules in an on-line mode will see the requirement for the database to be integrated with the control system being modelled in order to allow for downloading of current status data of the system, giving truly on-line facility.

Model building

Fig. 5 shows the typical outline of a simulation project. Although this is not to scale, it indicates that the amount of time spent in actually writing the model is relatively small (a maximum of 30%) in comparison to the

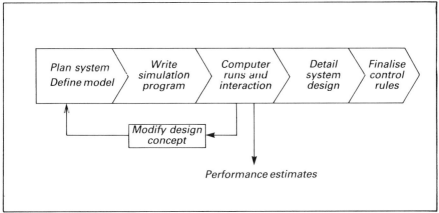

Fig. 5 Outline of a typical simulation project

total effort of the project. It is therefore important not to get this out of perspective when looking at model building techniques. There are three normal approaches to model building.

Modular packages. These are generally made up of a library of statements that require certain parameters to be set. These statements are the building blocks for all the visual, interactive simulation features. A knowledge of FORTRAN is required to generate the logic and decision rules required in any particular system.

Packages with a code generator. In order to make model building faster some systems include a code generator. This is a user-friendly front end system that allows the user to define all the areas of the model in either an English-like question-and-answer form or from a menu. The level of development of this type of system currently means that, in the case of more complex systems where there are many forms of decision rules and priorities, the model builder is still required to write FORTRAN code. The advantage therefore is somewhat limited in many applications.

Data-driven packages. These packages require no knowledge of specialist languages on the part of the model builder. The model is built up using the data of the proposed system (i.e. number of machines, performance, etc.), which is fed into an already validated model. The disadvantage of this is that the validated model generally has one particular area of applicability, such as FMS, tool management, distribution, etc., and is therefore limited to use in these areas. Packages are available that allow additional logic to be built in, although this requires an intimate knowledge of the assumption on which the validated model was built.

Modular packages are currently the most popular type with those users who may require to model several different types of application, such as a batch machine shop, an FMS, conveyor systems, etc. There are, however, some packages which combine features from all the above approaches. These will trade off flexibility by relating to a small set of applications (e.g. batch operation) for ease of use by containing built-in subroutines (e.g. for contention of conveyors).

Output methods

Apart from a dynamic graphical display it is also important that the system chosen should include a good standard range of statistical output reports, and various means of displaying them (graphs, histograms, etc.), either on the screen or on hard copy. It may also be desirable to link a colour printer or camera to the system at some stage, and the possibility of doing this should be checked.

Hardware

The nature of simulation is such that relatively powerful computers are required to carry out this highly iterative process fast enough to allow the study of system performance over extended simulated periods within short real-time periods. The additional overhead of graphics processing requires a special graphics terminal in the case of visual interactive systems. This is then linked to the host computer, which is generally a super-micro or a minicomputer. The advent of more powerful and faster microcomputers has spawned systems which can cut development time at the expense of limited graphics and interaction and slower simulations. These are adequate for strategic purposes but the limitations can be critical if detailed explanation of a system's performance is required. A comparison of the advantages/ disadvantages of these systems is given in Table 1.

A check list for simulation systems

The sections above have covered briefly some of the more important features to consider when choosing a simulation system. It is impossible to cover all the factors, but the following gives some additional points to consider in the form of a check list, which it is hoped will prompt

Table 1   Advantages/disadvantages of super-micros and minis as host computer

Minis	$\pm$	Micros	$\pm$
More powerful faster simulation runs	+	Very large models can run quite slowly	−
Not portable (physically)	−	Portable (physically)	+
Multi-user possible	+	Only single user	−
Can suffer from batch processing overheads	−	Truly interactive with single user	+
More expensive	−	Cheaper	+
Longer to develop	−	Quick to develop	+

potential purchasers to question the vendors deeply when choosing their system:

- What quality graphics?
- How large a system be portrayed graphically?
- Is there a serious limit to the size and complexity of the model that can be built?
- Are the graphics animated in simulated time or do they 'play back' after simulation runs?
- Is there a high level of interaction available during model building and model runs?
- How are models built with this package? How skilled? How fast? What limitations to complex logic building are there?
- Is the software limited to particular kinds of operational system?
- Is there a relational database package?
- How quickly can models be changed and run?
- How portable is the software? Is the hardware suggested of reputable type?
- What training is available?
- What is the documentation like? Clear, detailed enough?
- Will the vendor give the suport required? What future developments are planned?
- Can the package be used for applications other than simulation?

Using a system

Who should use it?

An important consideration in choosing a system is who should do the model building. This may well have a bearing on the type of system chosen or indeed whether it is felt necessary to take outside help. It is widely felt that in the case of the study and design of advanced manufacturing systems it is the manufacturing engineers that should carry out the simulation and not a 'simulation expert'. Although the initial overhead may be larger, ultimately the knowledge of the system gained through the process of design via simulation will benefit the manufacturing department greatly. In this way the expertise is retained intimately within the group most interested in the success of the proposed system.

What is the cost?

The cost of simulation is still a significant one, as pointed out earlier in the paper. It is made up of two elements: first the system cost (hardware and software), and secondly the cost of becoming expert in using it. Although it is difficult to generalise, the annual cost of the system, other overheads and the model builder will probably be in the region of £25,000 per annum.

If the potential purchaser is able to justify this expense by spreading the cost over a number of projects, it is sensible to purchase the system.

This has the advantage of creating and retaining the expertise within an organisation and will eventually pay dividends.

However, there are many companies that will perhaps only have one application suitable for simulation. The use of external consultants in this case gives the organisation the possibility of undertaking a simulation project together with an organisation which has the most up-to-date system and which has applied simulation techniques in a wide variety of applications.

If the potential purchaser is unsure of the applicability of simulation in solving his particular problem, it is possible to have a pilot model built at fairly low cost by external consultants. In this way both the applicability of simulation and of a particular simulation system can be tried out without heavy financial commitment.

Concluding remarks

Simulation is now a proven management tool in the planning and design of AMT. It gives an alternative basis to the traditional 'boardroom report' on which to evaluate large capital investments. Its use can be justified both by the direct savings it can yield and, probably more importantly, by the confidence it gives in the understanding and planning of advanced manufacturing systems. The cost is significant, however, and is the first question to consider in evaluating a system. The development of powerful visual interactive simulation systems that make use of animated colour graphics has had a enormous impact on the understanding and acceptance of simulation by manufacturing industry and it is now the manufacturing engineering departments that should and are taking advantage of this facility.

Several of the most important features of a system have been discussed. These are:

- Graphics.
- Interaction.
- Database management.

- Model building.
- Hardware.

While one option is to purchase a simulation facility, many companies will not have a sufficiently high demand on their system to justify the orginal cost or the cost of getting to know simulation techniques and the particular characteristics of a package. In this case the use of consultants as a partner in solving manufacturing problems enables an organisation to gain access to the most up-to-date system and knowledge. This is also a way of testing both the applicability of simulation techniques and of a particular simulation system to the specific problems encountered in an organisation.

Whichever method of using simulation and whatever system are considered as best, the certain message now is that simulation is here to stay and that if companies wish to compete in increasingly challenging markets it must be considered as a vital management tool that they can not afford to ignore.

PRACTICAL EXPERIENCE CONTRASTING CONVENTIONAL MODELLING AND DATA-DRIVEN VISUAL INTERACTIVE SIMULATION TECHNIQUES

S.R. Hill and M.A.M. Rogers
ICI plc, UK

Case studies for conventional and data-driven visual interactive
simulation systems are described for applications in the chemical
industry. Useful data on effort, skill requirements, code size, etc., are
given and comparisons of the two techniques are made.

ICI, one of the world's largest chemical manufacturers, has used simulation techniques for many years for modelling its batch plants and its mixed batch and continuous plants.

Chemical batch plants are subject to an increasing rate of change in terms of new products, process improvement and operations management. Even without these developments plants can be very difficult to operate at high throughput due to the many possibilities for product routing, the range of products produced, the mix of manual and automatic operations and contention for common services.

Incorrect decisions in the design, operation and modification of chemical batch plants can lead to wasted capital outlay, poor output, lost market opportunities and deteriorating customer relationships. For these reasons ICI has developed a range of decision support system products which are applicable throughout the life cycle of batch plants[1]. Two of these products, VUSIM[2] and PROPHET[3], are used to construct simulations.

This paper recounts ICI's experience of using both conventional (VUSIM) and 'data-driven' (PROPHET) visual interactive simulation techniques with reference to two specific models.

A categorisation of simulation software packages
Simulation packages fall into three categories as follows.

Non interactive

Here the package provides a simulation mechanism which the user customises to his application by adding his own routines written in a language such as FORTRAN. The program is run in computer 'batch mode' and the output is typically lineprinter hard copy. This type of package is usually very versatile in that it can be used for most application areas. However, due to the nature of the output, validation of the model can be difficult and interpretation of the results has to be left to the application analyst. The decision-maker, who may be responsible for a large capital investment, has to rely on the interpretation of the analyst – he cannot observe the running of the model for himself.

Conventional interactive colour graphic

Here the package provides a simulation mechanism which is usually event based, interactive service subprograms for debugging, snapshot dumping, fault handling, service subroutines and display facilities. Output is typically in the form of a mimic display which is backed up by histograms, time series and tabular data. Top-class packages, such as VUSIM[4], run on a variety of machines (e.g. VAX, IBM PC XT/AT) and can drive a number of terminal types. These packages are very versatile and can be used for any problem which can be described in terms of discrete events. As with the previous category, the package is customised to the application by the addition of analyst-written subroutines.

'Data-driven' interactive colour graphic

This type of system is also described as 'generalised' or 'parametrised' or 'fourth generation'. The aim of this type of package is to reduce the time taken to produce and modify a simulation. This is done by reducing or removing the need to program in a language such as FORTRAN. Instead the model is 'programmed' by either constructing a special data file which contains all the model parameters or, as is the case with PROPHET[5], by means of conversational dialogue and a powerful high-level language.

A VUSIM model used to double production

The plant

The plant consists of three identical lines each of which has six stages. There are 13 vessels in each line. This is depicted in Fig. 1.

Stages 1 to 5 are production stages. Stages 1, 2, 4 and 5 are batch operations while stage 3 is a continuous operation. Stage 6 is the packing stage, which is a continuous operation.

The plant operates a five-day week, two-shift system.

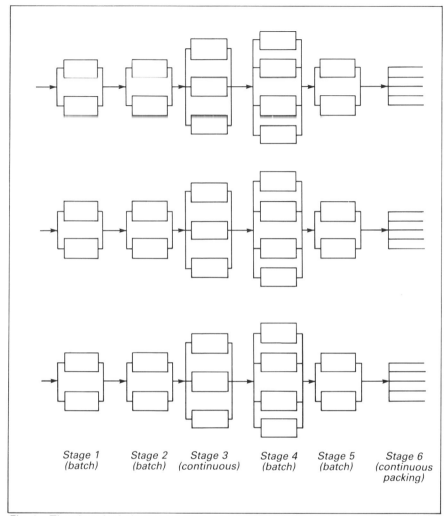

Fig. 1   The chemical plant modelled using VUSIM

The products

There are around 15 products, with new products or variants appearing every few months. They are produced in runs which last from a few days to two months or more.

The process

The process detail is different for each product but is summarised by the cycle diagram[6] shown in Fig. 2.

There are complications due to, for example, time windows. That is, some vessels cannot remain full overnight or at weekends and the laboratories used in Stage 4 are only open for the dayshift. Some products require two of the three production lines.

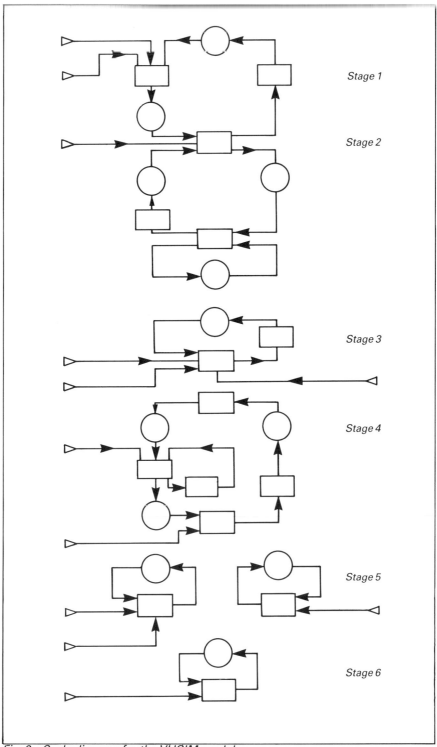

Fig. 2   Cycle diagram for the VUSIM model

The problem

Marketing indicated that double the then present output could be sold in the coming year. Stage 3 was considered the bottleneck and it was expected that more than £1 million would need to be spent to increase capacity here.

Before any capital expenditure was committed a VUSIM model was constructed.

The model

VUSIM is a discrete-event simulation package which is customised by the addition of FORTRAN subroutines. It can be used to model any process which can be described in terms of discrete events. The discrete events in the case of the application described here are essentially the starting and stopping of the activities shown in Fig. 2. If the code is constructed modularly (as it was here) there will be one subroutine per event. The size of the subroutines depends on the complexity of the process.

The application described here is large and complex, and hence the following statistics on the model represent something of an upper limit for a VUSIM model:

- Analysis, coding and testing effort: 80 man-days
- Total lines of executable FORTRAN code: 4000
- Analysis of code function: 50% event code
  30% data editing code
  20% display code
- Number of events: 40
- Average size of event subroutine in executable lines of code: 44

Before a model of this type is constructed a decision must be made concerning what items in the model are to be variable and what items are to be fixed. In this model the vessel capacities, plant rates, working hours and product recipes are variable. Data editing code must be provided for all variable items. This code allows plant rates, etc., to be changed via interactive dialogue.

A decision must also be made concerning the form of the displays and measures of performance. In this model the main display is a colour-coded mimic representation of the plant and the measures of performance are plant utilisation and output figures presented in the form of histograms and time-series graphs.

How the model was used

A modelling project of this kind needs very close collaboration with the plant management if it is to be successful. The Plant Manager was involved in the construction and validation stages and was also in the 'experimentation' with the final model. This was easy and trouble free due to the 'friendly' nature of the interactive dialogue and plant mimic diagram.

The results

Trials of all products under different operating scenarios showed that contrary to prior belief Stage 3 was not the bottleneck. It soon became clear that much of the time on one group of vessels was spent waiting for laboratory tests, largely because production was on two shifts and laboratory cover was limited to normal day hours. A simulation test with the laboratories on two shifts showed a doubling of production on some products.

With the laboratories on two shifts the model showed that the packing lines could not cope with the extra output. The model was used to determine the rating for a new packing line, the cost of which was about one-third the amount that was originally expected to be spent on the plant.

The sanction for capital expenditure and extension of the laboratory working hours was relatively straightforward due to the fact that the works management had seen the model and could easily understand its mimic and data displays.

A PROPHET model used for process analysis

The plant

The plant is mixed batch and continuous with five stages. This is depicted in Fig. 3. Stages 1 to 3 are production stages. Stage 1 is a batch process which feeds into one of four large portable containers. The

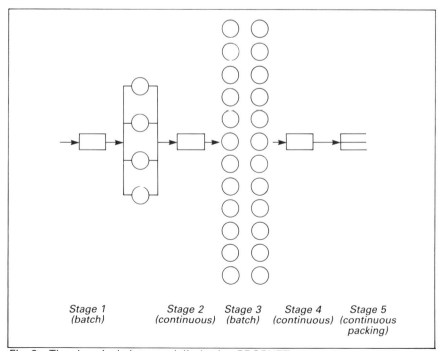

Stage 1 (batch) Stage 2 (continuous) Stage 3 (batch) Stage 4 (continuous) Stage 5 (continuous packing)

Fig. 3   The chemical plant modelled using PROPHET

contents of the containers are transferred into Stage 2, which is a continuous process. The output from Stage 2 goes into Stage 3 (i.e. small portable containers – 26 in total), where it is left for a time to 'cure'. From here the product goes into a vessel (Stage 4) and then on to packing lines (Stage 5).

The product

Only one product is produced on the plant.

The process

The process is summarised by the cycle diagram shown in Fig. 4. There are various conditions for starting and stopping the stages in the process (e.g. the process stops if no small containers are available).

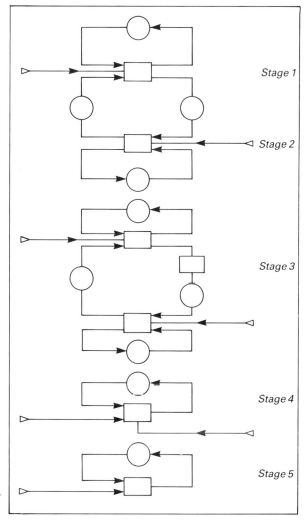

Fig. 4   Cycle diagram for the PROPHET model

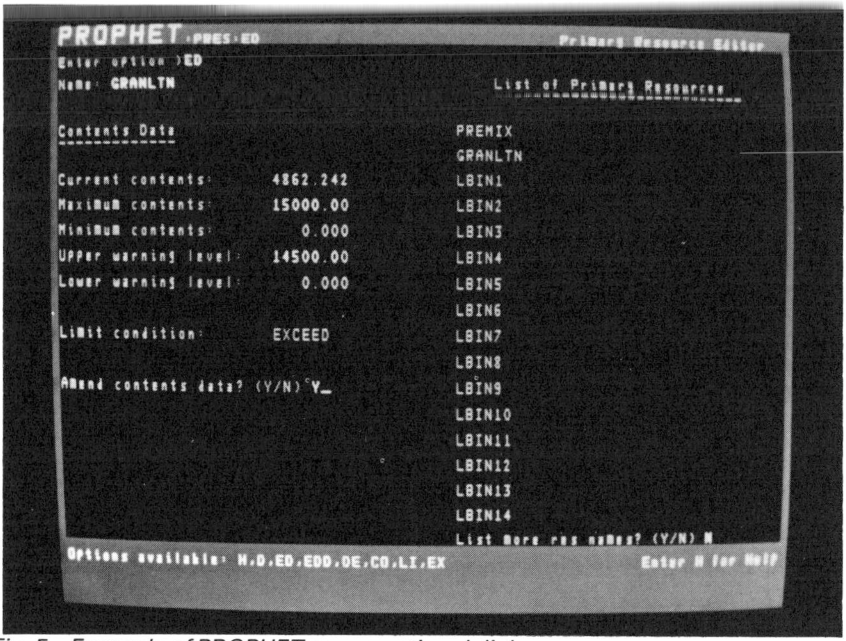

Fig. 5   Example of PROPHET conversational dialogue

The problem

The Productivity Services Officer wished to know where the bottleneck was with a view to improving the output of the plant.

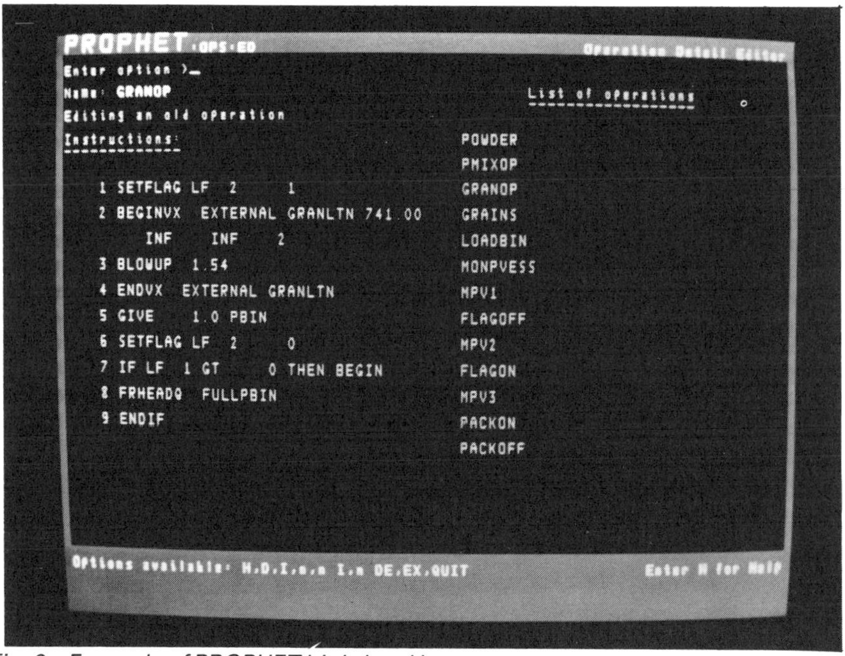

Fig. 6   Example of PROPHET high-level language

The model

PROPHET is a process-orientated simulation system which has been designed to enable simulations to be constructed in a matter of days by a wider range of people than is the case with conventional packages. It does this by removing the need to program in FORTRAN. Instead the model is created via a conversational dialogue and a high-level language. Figs 5 and 6 show examples of this (NB the originals are in colour).

The application described here is a typical one. Some statistics for this model are:

* Analysis, coding and testing effort: 2 man-days
* Total lines of high-level code: 43

PROPHET provides high-quality configurable displays in the form of mimic diagrams, bar charts, time series, histograms and tabular data. Examples are shown in Figs. 7 and 8 (NB the originals are in colour).

The results

A thorough understanding of the plant dynamics was gained by the Productivity Services Officer. The model was shown to the Plant Manager and will form the basis for future work in improving the efficiency of the process.

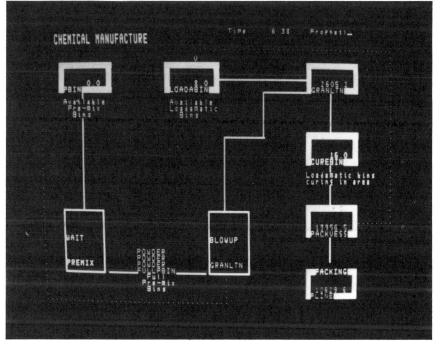

Fig. 7   Example of PROPHET mimic diagram

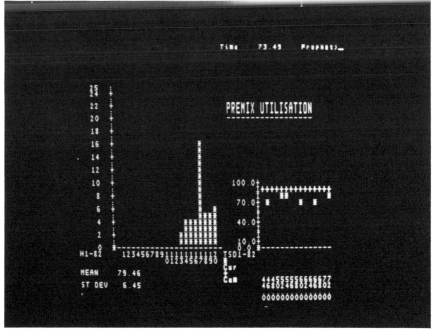

Fig. 8   Example of PROPHET time series and histogram

A comparison of the two modelling techniques

Project costs

Simulation models are used for strategic decision-making. In manufacturing industries this is usually in the form of process analysis – how many reactor vessels, the best product mix, which process, etc. This decision-making can be at the design, modification or operation stage and can involve small or large sums of money and small or large elements of risk. Like any other business undertaking a simulation should have an associated cost-benefit justification.

In the first example discussed in this paper the decision involved capital expenditure of the order of £1 million. The simulation costs were between 1 and 2% of this. Given the complexity of the plant processes and the implied uncertainties the cost of the simulation was amply justified.

In the second example discussed in this paper the cost of the simulation was two man-days of effort. The benefit was a far better understanding of the process dynamics and an easy-to-use model with which to experiment. At this price every plant should have one!

As a guide to those costing simulation projects, ICI would expect project costs to scale according to the complexity of the cycle diagram (see Figs. 2 and 4). Note that PROPHET models cost approximately one-tenth of the equivalent conventional (VUSIM-type) model costs.

The costs discussed here do not include once-off costs for software and hardware. Information on this is available from the authors.

Skill requirements

Building a useful simulation model requires skills in three areas:

(i)   Modelling.
(ii) Use of the simulation package.
(iii) The relevant process technology.

For conventional simulations it is quite common for skills (i) and (ii) to reside with the OR specialists and for skill (iii) to reside with his client, who may for example be a process technologist. Acquiring FORTRAN programming skills is often a great hurdle for many people, with the result that few process technologists learn how to construct models.

This division of skills can lead to mistakes and inefficiency. There may also be a shortage of skilled OR specialists.

PROPHET-type packages make it easy for non-OR specialists to construct models. They do this by clearly defining the construction procedure and by drastically reducing the amount of coding and computer knowledge required. It is important that non-OR specialists understand the basics of modelling (e.g. what is relevant to the model and how to treat stochastics).

As a guide for training non-OR specialists, ICI would expect a conventional visual interactive simulation package to require a five-day training course, followed by a learning curve taking say one month to reach a reasonable level. A PROPHET-type package would require a two-day training course followed by a one-week learning curve to reach a reasonable level.

Timescales

In a business environment where there is a high turnover in products and where a competitive advantage can be gained by rapid action, it is sometimes very important that simulation projects are completed quickly. The speed with which PROPHET models can be built could mean the difference between having simulation results ready in time and not having them.

Level of detail achievable

FORTRAN-based simulation packages such as VUSIM are capable of representing the problem area to any level of detail.

Data-driven packages such as PROPHET are aimed at being applicable to say 95% of the applications in their problem area and are by their nature less versatile than FORTRAN-only packages. A common solution to this problem is to allow small amounts of FORTRAN code to be 'plugged in' to data-driven packages. This must be done in a controlled manner to ensure integrity of the model and specific utility routines should be provided to aid the programmer.

The PROPHET models constructed to date have not used extra FORTRAN code.

Visual appeal and 'sellability'

The effort of constructing and running a simulation is wasted if the results cannot be 'sold' to the decision-maker. This 'sale' will be much easier if the decision-maker can see the model running on a mimic or bar chart and can 'play' with it to ensure, to his own satisfaction, that it represents the plant concerned and that the results, therefore, are trustworthy. This is one of the main benefits of visual interactive simulation.

The VUSIM model discussed in this paper has a plant mimic as its main display and has been demonstrated to the top management levels in the business concerned.

While the form of the display for FORTRAN-based packages is wide open for the analyst to design as he pleases, data-driven packages have several standard displays which are 'customised' to suit the application. PROPHET has mimic diagrams, bar charts, histograms, time series and tabular data displays. All can be configured and mixed and embellished with static text and pictures.

Problem areas that can be tackled

FORTRAN-based discrete-event simulation packages are completely general in that they can be used to tackle any problem which can be represented in terms of discrete events. This includes chemical batch plant processes, continuous operations which can be modelled by start-stop events, and applications such as telephone exchanges where there may be large numbers of entities involved. The package discussed in this paper, VUSIM, has been used for applications as diverse as chemical plant, computer assembly lines, plant operator training, and a site security system (electronic gates, etc.).

Data-driven packages, on the other hand, are aimed at specific application areas. For example, PROPHET is aimed at batch and mixed batch and continuous process type applications and at applications that can be represented on a cycle diagram or bar chart. Examples of the use of PROPHET include chemical plant, a dyehouse and a shipping model to determine dock capacity.

Concluding remarks

Both conventional and data-driven interactive simulation systems have their strengths and weaknesses. VUSIM-type systems can be used for a wide range of applications. PROPHET-type systems are very quick to use. Both types of system are put to use in ICI with great benefit.

References

[1] Rogers, M.A.M. 1985. Lifecycle simulation for design, modification and management control of batch processes. In, *Proc. 1st Int. Conf. on Simulation in Manufacturing*, pp. 355-358. IFS (Publications) Ltd, Bedford, UK.

[2] VUSIM Sales Literature, ICI Americas Inc., Wilmington DE 19897, USA, or ICI CMSD, P.O. BOX 11, Runcorn, UK.

[3] PROPHET Sales Literature, ICI Americas Inc., Wilmington DE 19897, USA, or ICI CMSD, P.O. BOX 11, Runcorn, UK.

[4] Hill, S.R. 1985. *The VUSIM Visual Discrete Event Simulation Package*, ICI Report Number IC 06436. ICI Americas Inc., Wilmington DE 19897, USA, or ICI CMSD, P.O. BOX 11, Runcorn, UK.

[5] Hill, S R 1986, PROPHET Overview, ICI Report Number IC 06553. ICI Americas Inc., Wilmington DE 19897, USA, or ICI CMSD, P.O. BOX 11, Runcorn, UK.

[6] Hutchinson, G.K. 1975. Introduction to the use of activity cycles as a basis for system's decomposition and simulation. *SIMULETTER*, 7(1): 15, 20.

7

Expert Simulation Environments

A manufacturing simulation project will usually involve experts from several disciplines. The emerging field of expert systems/artificial intelligence contain methods of coding expert information in the form of knowledge bases. This section describes how expert systems may be used to control a manufacturing simulation and how knowledge may be coded into an FMS advisory simulation environment.

A KNOWLEDGE-BASED SYSTEM FOR SIMULATION AND CONTROL OF FMS

D. Ben-Arieh
AT & T Bell Laboratories, USA

In a modern manufacturing facility the available flexibility introduces another degree of complexity in decision-making. The requirement for a real-time control mechanism generates a need for an on-line simulation system that can simulate the complicated scheduling, routing and similar decisions in the system. A knowledge-based simulation and control system written in PROLOG is described. The manufacturing system under investigation is a combination of flexible manufacturing cells that feed an automated assembly station.

Modern flexible manufacturing systems (FMS) are complex systems with a high degree of flexibility. Control of such a system is done hierarchically in various levels[1]. The middle level performs the scheduling of parts and tools in the system. This task has to be done in real time and is too complex to allow a steady-state optimal solution to be implemented. One way to control such a complex system is to use a knowledge-based controller that has the capability to solve in real time the various problems in the manufacturing environment[2,3]. In order to verify the knowledge and tune the controller there is a need to combine it with a simulation system.

This paper discusses the need for a new simulation and control environment for FMS. The knowledge-based simulation environment is presented, and an existing system is shown. This system simulates the FMS in order to tune its scheduling and routing algorithms. The knowledge base can then be used to control the FMS. The system is based on the 'KBRS'[4] and is written in C-PROLOG.

Knowledge-based simulation versus network simulation

In analysis of automated manufacturing systems one uses tools on three levels of sophistication (excluding analytical models). The first and simplest includes the modelling tools like the various automata types and Petri net family of models. The second level contains the simulation tools. The third level comprises the automated planning models.

A knowledge-based simulation is a modelling and analysis tool, on a level between the simulation and the planning models.

Network simulation environment

There are various ways to describe the network simulation environment. According to Reilly et al.[5], such an environment is made up of four parts: simulation builder, record keeper, model executor and result analyser.

An approach from general systems theory is found elsewhere[6]. In this approach the simulation model is composed of the model, a generator of input segments, an acceptor that accepts the right input, and a transducer that performs the summary mapping.

From a structural point of view, a network simulation is composed of three elements:

- *The modelling part.* This part contains the knowledge about the system in terms of the interaction between its various components. It describes the flow of the entities, conditional branching, resources available and their consumers.

- *The database.* This part contains both the attributes of the system that do not change with time (like service time, number of servers at a station, branching probability, stopping time) and those that change dynamically with the system state. It keeps track of the entities that flow in the system, and the state of the system.

- *The output management system.* This part is usually a general statistics collection routine that calculates the mean waiting time in a queue, queue length and other quantities assumed useful.

Usually in a network simulation language the three functions are not separable, and the user has no control over some of them.

Knowledge-based simulation

The use of a knowledge base for simulation is a new concept that has been introduced in the early 1980s. One of the early systems was the ROSS system[7], used to create a very large scale simulation of military air battles. This system was written in an object-orientated language called Director. A latter system was the KBS[2], which used the SRL language. This system allowed different levels of abstraction, and some completeness and consistency checking of the model. Doukidis and Paul[8] discuss the involvement of expert systems in simulation. They use activity cycle diagrams as a modelling tool and two types of knowledge: facts of the domain and heuristic knowledge.

Recently PROLOG and some of its dialects have been used as simulation and modelling tools. Some of PROLOG's capabilities – symbol manipulation, pattern matching, intelligent database management and searching – are useful for these types of tasks. The dialects

T-PROLOG and TS-PROLOG[9] are a good example of the usefulness of this language for modelling and simulation. LISP has its own merits as a simulation tool, as described by Sergio et al.[10]. A simulation of a complicated robot-controller using LISP[11] shows the flexibility of such a modelling environment to represent different types of systems, answer 'what-if' type questions, and allow different levels of abstractions and functional expansion.

Some of the differences between knowledge-based simulation and traditional network simulation are[12]:

* Regular simulation is primarily numeric, while knowledge-based simulation involves many symbolic processes.
* Regular simulation is algorithmic with explicit steps, while the other uses pattern-invoked searches.
* The information and control are integrated in regular simulation, whilst in a knowledge-based system they are distinct parts.
* Knowledge-based simulation can include the usually external steps of testing input data, designing the experiment, etc.

This knowledge-based simulation environment is composed of four elements:

* Static database.
* Dynamic database.
* Modelling knowledge.
* Simulation driver.

This structure is described in Fig. 1.

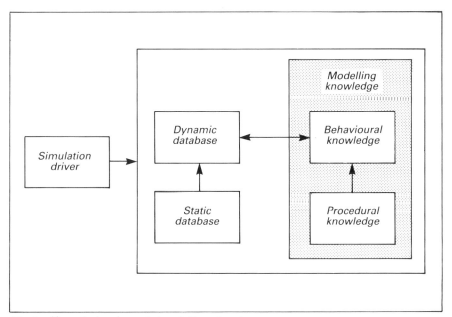

Fig. 1   Knowledge-based simulation structure

The advantages of a knowledge-based environment over a network simulation environment are:

- The four modules of the simulation are distinct and reachable by the user. The user can modify any of them without affecting the others. This increases the modularity of the system, its understandability and uniform structure.
- The user can specify the output required for decision-making or understanding of the system.
- The system comprises various layers with increasing detail, and the user can use the desired level of detail for the simulation.
- Knowledge-based simulation can be used for planning purposes. It can try to satisfy some goals, or to find a reverse chain of actions that will generate the goals. An example is the simulation of a bank robbery written by Futo and Szeredi using T-PROLOG[9].
- The system has an enhanced modelling power. It can simulate very complicated decision mechanisms, and can even simulate into the future to check a chosen policy.
- With is modular structure the simulation environment can be used for control purposes. This is done by disconnecting the simulation driver from the system and connecting the real world and the event generator.

The knowledge-based controller structure

The knowledge-based controller is composed of four main parts: static database, dynamic database, modelling knowledge, and simulation driver. The system has two levels of knowledge: the first decides upon the immediate solution to the routing problem, while the second observes the behaviour of the system and infers the adequate rules and knowledge that are required to achieve the best performance from the first level.

The controller/simulator is a production system, which means that the knowledge is represented in sets of IF-THEN rules[13]. In this system two types of knowledge exist: production knowledge (rules), and procedural knowledge which is responsible for the algorithmic part of the routing process. The production type of knowledge is presented in predicate form using PROLOG.

The static database

The static database is the source of the information that describes the process and the simulator, and does not change with time. Examples are the process times for the different operations on the various machines, the structure of the product, the initial and final conditions for the simulation, and any other values known in advance. The database is represented as facts in PROLOG. Examples of this part are the predicates:

buffer _ size (A, Size)

where A is a buffer name and Size can be instantiated,

process _ machine (1,[1,3,5,6])

This predicate specifies the processes that can be done on machine 1. The durations of the various processes are also registered in the static database, in the following way:

part _ time (a1,[process(1,[time(1,4),time(2,6)]),
        process (8,[time(3,7),time(4,9),time(5,10)]),
        process(9,[time(2,4),time(3,6),time(5,8)])])

This structure of nested lists stores the part name, the processes it needs to take, and their times on each feasible machine. The structure of an assembled component is stored as:

part _ of (a1,wheel)
part _ of (wheel,axle)
assembly (wheel,[quantity(a1,1),quantity(a2,1),quantity(a3,3)])

In this manner the entire product structure can be represented.

The dynamic database

This part of the knowledge-based controller stores the data that represent the state of the system. It contains the queue sizes and contents, the status of the machines, the current processes performed on each part, and the tools currently available in each station. Examples for this part are:

queue(1,[wait _ time(a2,11),wait _ time(b2,4),wait _ time(d1,1)])

This predicate specifies the parts in queue number 1, and their time in the queue at that moment.

status(machine _ 1,up)

which represents that fact that machine 1 is in operating condition.

current _ process(a1,9,machine _ 1)

This predicate states that part a1 is currently on machine 1 for process number 9.

In addition to the predicates that dynamically describe the system, there are the predicates that collect the information for statistical analysis by the simulation driver. These predicates are triggered by the output requirements of the user, and accumulate data for end-of-run manipulation. Examples are current busy time of the machines, current queue length, accumulated time of the queue in this length, waiting time accumulated so far at each station, number of parts waiting at the stations, and similar data. Also, this information is formatted in predicates and is updated by the system in run time. Examples of these data items are:

accum _ busy _ time(machine△1,X)
part _ waiting _ time(a1,X) (for part-based statistics)
queue _ length(machine _ 1,Total _ time,Last _ time _ updated)

The dynamic database is responsible for all time-dependent data that both keep track of the system states and maintain data for analysis.

The modelling knowledge

This part is the main component of the simulator/controller in the sense that in it all the knowledge about the model of the system is formatted and the response of the system is described. The modelling knowledge is composed of two parts: behavioural knowledge and procedural knowledge. The behavioural knowledge describes the behaviour of the system in terms of actions that cause other actions to take place. This part is responsible for the generation of events in the system and for performing the events in the event list. It is written in terms of predicate logic and production rules using PROLOG. In order to control the rule selection, the PROLOG matching property is used. Since several rules may have the same left-hand side, a conflict resolution mechanism is required. This is achieved by choosing the rules according to their order in the database, and partly using a 'context-limiting' strategy. Therefore, in order to make a rule more favourable it is possible to move it upwards in the list, and it is chosen by considering its context, which is mentioned immediately on the right-hand side.

The other part of the knowledge, the procedural knowledge, quantifies the routing decisions and helps to choose the preferred one, according to the real-time condition of the FMS.

The two knowledge parts have to cooperate in defining the problem and supplying the information from the databases. In the system considered here, a simple forward simulation to support earlier decisions was used as a part of the decision procedure.

The simulation driver and the simulation

The simulation described here has five main events: next _ arrival, process _ finish, end _ assembly, machine _ failure and machine _ repair. Each event triggers hidden lower level events, except machine _ failure, which directly schedules the machine _ repair event. Examples of lower level events are: add _ to _ queue(Q), remove _ from _ queue(Q), choose _ next _ machine, and so on. Each one of the events is further decomposed from lower level events, until the lowest level events are reached and directly change the data structures. Each level has a set of primitive events that the modeller can apply to, thus creating the various levels of detail determined by the user needs. In addition to the primitive events that the user can use to change the dynamic behaviour of the system, the databases can be modified in order to create a change. The database is considered to be on the highest level, and is available to any kind of user. Therefore, unless a substantial change in the system behaviour is required (like the hidden assumptions), a change in the database can initiate many of the desired investigations.

The simulation driver is responsible for the simulation of the model. Basically this part is a small one, and is made up of the event file and the simulation clock. Since the behaviour of the system is defined by the modelling knowledge and the databases, all that is left is the event file, which stores the list of predicted events with their times. The events are generated by the modelling knowledge part, and therefore the simulation driver is only the manager of the system. This gives the simulation the capability of various sorting procedures for the event list, as well as any type of tie-breaking desired by the user. This part is also transparent to the unsophisticated user and can be disconnected at will, to change the system into a control system.

The event list is represented by the list:

$$event \_ list([Event \_ time, Event \_ code, Attribute \_ list)$$

This list specifies the time at which the specific event has to occur (Event _ time), its name (Event _ code), and the specific list of attributes that describe the event. For example, event next _ arrival has one attribute which describes the part type that arrives. Event part _ finish has two attributes: the part name and the machine number on which the process is finished. In addition, there are predicates that generate random numbers, sort the event list, remove the old events and add the new ones.

The modelling knowledge can be triggered by the events simulated or by the real system. For example, the event of finishing a process on a machine can be simulated by the simulation driver or sensed in the real system. In this way the system can first simulate various scheduling algorithms or other control methods. When the behaviour of the system is satisfactory, this environment can be interfaced with the real system. The same response will control the real system and the simulated one. In order to change the system from simulation to control phase, the user needs to specify this at the menu-driven interface, at the beginning of the run, and to connect the controller to the correct interface with the manufacturing system.

The manufacturing system

The simulation-control environment is used to control an FMS facility composed of a machining cell that feeds an automated assembly station. The manufacturing system is described in Fig. 2.

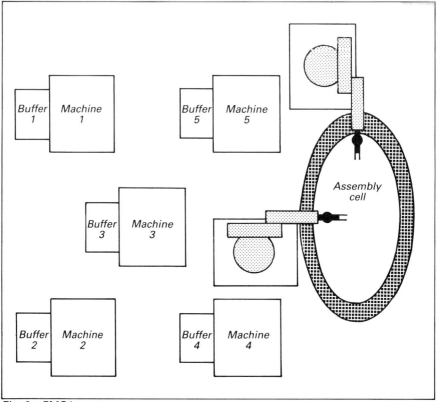

Fig. 2   FMS layout

The machining cell is composed of five DNC machines, and serves 14 components, nine of which make up an end product. The production of each component comprises up to five processes. Each process can be performed by almost all the machines, with different degrees of efficiency. Therefore there is a need for a real-time controller to route the components in order to achieve optimal production. Some of the additional factors that need to be considered are failures of the machines, and the ability of each assembly station to perform any assembly operation for which the components are available.

Concluding remarks

This paper has described a new approach to controlling and simulating a complicated manufacturing system. The controller uses a knowledge base in order to make decisions, and decisions are simulated in order to evaluate and improve them. Improvement of the decisions is done by changing the knowledge base (changing the procedural knowledge and the databases), tuning the algorithms used in the decision-making process, and adding a layer of second-level knowledge (meta-knowledge) that chooses the right approach according to the current conditions of the system (based on past experience).

The system described here was run on a VAX 11/780 machine under a UNIX operating system. This control method, when compared with other common static scheduling routines, gave excellent results. It was able to produce, when simulated, four times more assembled products than the most sophisticated dispatching rule tried. The difference was even bigger when the machines had a significant failure rate.

References

[1] McLean, C.R., Bloom, H.M. and Hopp, T.H. 1982. The virtual manufacturing cell. In, *Proc. 4th IFAC/IFIP*, Gaithersburg, MD, USA, October 1982.

[2] Fox, M.S., Allen, B. and Strohm, G. 1982. Job-shop scheduling: An investigation in constraint directed reasoning. In, *Proc. NCAI*, pp. 155-158.

[3] Nof, S.Y. 1983. An expert system for planning/replanning programmable production facilities. In, *Proc. ICPR*, Windsor, Canada.

[4] Ben-Arieh, D. 1985. Knowledge based control system for automated production and assembly. Ph.D. Dissertation, Purdue University, USA.

[5] Reilly, K.D., Jones, W.T. and Dey, P. 1985. The simulation environment concept: Artificial intelligence perspective. In, Holmes (Ed.), *Artificial Intelligence and Simulation*.

[6] Zeigler, B.P. 1985. *Multifacetted Modeling and Discrete Event Simulation*. Academic Press, New York.

[7] Klahr, P. and Faught, W.S. 1980. Knowledge based simulation. In, *Proc. 1st Annual National Conf. on AI*, Stanford, CA, USA.

[8] Doukidis, G.I. and Paul, R.J. 1985. Research into expert systems to aid simulation model formulation. *J. Opt. Res. Soc.*, 36(4): 319-325.

[9] Futo, I. and Szeredi, J. 1984. System simulation and cooperative problem solving on a PROLOG basis. In, Campbell, J.A. (Ed.) *Implementation of PROLOG*. Ellis Horwood, Chichester, UK.

[10] Sergio, R.M., Talavage, J.J. and Ben-Arieh, D. 1985. Towards a knowledge based network simulation environment. In, *Winter Computer Simulation Conf.*

[11] Ko, H.P. and Wheeler, J.E. 1983. A computer simulation methodology by LISP and SANS. In, *Summer Computer Simulation Conf.*, pp. 40-47.

[12] Shannon, R.E. and Mayer, R. 1985. Expert systems and simulation. *Simulation*, 44(6): 275-284.

[13] Davis, R. and King, J. 1985. An overview of production systems. *Machine Intelligence*, 8: 300-332.

AI-BASED SIMULATION OF ADVANCED MANUFACTURING SYSTEMS

J. Shivnan
Digital Equipment International BV, Eire
and
J. Browne
University College Galway, Eire

The applicability of artificial intelligence to simulation in manufacturing is illustrated through the development of a control system for a small production facility. The simulation of this facility is described, together with the method of applying AI-based simulation to scheduling and real-time control within production activity control. Conventional simulation languages and AI - based simulation are also briefly compared.

Simulation has been accepted as an essential tool in tackling large-scale, complex problems in manufacturing systems. Within this domain three roles can be identified for simulation[1]:

- Design and analysis.
- Scheduling, particularly in automated systems.
- Real-time on-line control.

The importance of these activities in computer–integrated manufacture (CIM) has led to increased attention being devoted to artificial intelligence (AI) by the industrial and scientific communities. Thus the simulation community is not alone in seeking to explore the practicality of this technology for their discipline and several papers have recently appeared on the subject of AI and simulation[1-5]. This paper seeks to demonstrate the potential in using AI-based simulation in a CIM environment.

Influence of AI in simulation

Both AI and simulation are multi-faceted disciplines and so there are several different but not distinct research areas which consider the influence of AI. These may be classified into three areas as follows.

Intelligent support tools for simulation

Much of the interest in this area centres on the potential for rule-based systems to assist in the area of model building, experiment planning and data analysis[6,7]. A rule-based system is a programming organisation based on data- or event-driven operations. An operation is a module or a procedure triggered by the occurrence of a given pattern in the data stored in a database; its activity possibly modifies the contents of the database, allowing the firing preconditions of other operations to become satisfied. Shannon[11], for instance, describes the nature of an integrated expert systems support environment for simulation.

Complementary use of both AI and simulation in modelling

There exists an important relation between simulation and AI in that both are concerned with modelling: simulation in the field of objects and processes, and AI in the field of the human decision-making process. Consequently a joint application of both techniques to modelling a given situation results in a richer modelling capability.

Simulation capabilities in AI languages

Simulation capabilities have been implemented in various representation languages used by the AI community. These languages have either been languages which operate within a logic formalism, primarily PROLOG[5,8], or languages which use a frame-based formalism. In these languages knowledge, another important relation between AI and simulation, is organised into two categories:

- Declarative knowledge (facts).
- Procedural knowledge (how to).

It is within this area of simulation capability of AI languages that this paper concerns itself.

Simulation and CIM

In particular it is the capabilities of AI languages in a computer-integrated manufacturing environment that are of interest. CIM has been defined[9] as 'the integrated application of computer based automation and decision support systems to manage the total operation of the Manufacturing System, from product design through the manufacturing process itself and finally onto distribution and including Production & Inventory Management (P&IM) as well as Financial Resource Management'.

Within CIM and particularly in production activity control (PAC), a key node in CIM, simulation is seen as playing a major role as a decision support system from both the real-time control and scheduling points of view[10]. The notion of using simulation as a decision support system is important. Decision support systems imply the use of computers to[11]:

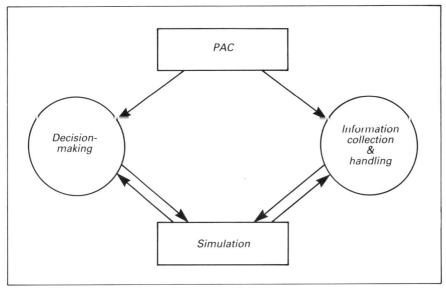

Fig. 1   PAC and simulation

- Assist decision-makers in their decision processes in semi-structured tasks.
- Support decision-makers, judgement.
- Improve effectiveness rather than efficiency of decision-making.

These activities represent the role of simulation in PAC as shown in Fig. 1.

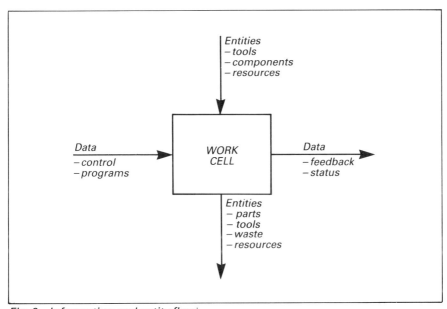

Fig. 2   Information and entity flow

With conventional simulation languages the emphasis is on the flow of physical objects through known processes. Thus the solution procedure is well understood. However, with regard to PAC, where data/information is the important aspect (see Fig. 2), the flow and effect of this information are not well understood. Due to the fact that decisions require consideration of a variety of alternatives and may involve uncertainty, this area is seen as a suitable application space for AI.

Knowledge within the PAC domain has various forms. Sections of this knowledge are heuristic rules of inference so that a rule-based system is an appropriate representation. However, domain data in the form of declarative knowledge (facts) may best be represented by frames. The deciding factor in choosing a representation is a question of economy of labour provided that the chosen scheme represents the required amount of knowledge to generate the appropriate behaviour. Consequently, OPS5 is the language used in the work described in this paper. However, it is recognised that a hybrid scheme of rule and frames would be more suitable.

OPS5

OPS5 is a rule-based language implemented in either BLISS or LISP. A rule-based system is composed of three elements:

- A rule base (OPS5 – production memory).
- A world model (OPS5 – working memory).
- A pattern matching and conflict resolution mechanism.

The rule base is a collection of knowledge organised in rules also described as productions. These productions have a structure as follows:

IF   condition <1>
                      condition <2>
                           :

THEN   action <a>
                       action <b>
                           :

Typically these rules are heuristic in nature and reflect the experience of the expert in his/her domain.

The world model stores all the date relevant to a problem. The data structures employed are simple, such as Entity, Attribute(s), Value(s). Thus, for instance, in OPS5 the following is possible:

(Machine	^Location	Insertion _ Area	^Type	DIP _ 374)
<Entity>	<Attribute>	<Value>	<Att..>	<Val..>

Given a rule base and a world model, a rule-based system must compare the rule base with the world model and determine which rules are applicable. A given rule may apply to more than one state of the world model. Likewise, more than one rule may apply to one particular state of the world model. (A state of the world model is similar to the states in conventional simulation languages.) In the case of many rules for one state, a conflict set is produced. Decisions on which rule to apply can be made on the basis of chosen factors:

- Most recent information.
- Most specific information.
- Highest priority information.
- Randomly.

The chosen rule is then instantiated or fired and modifies the state of the world model. Once a rule is applied it is removed from the conflict set until a state of the world model exists where it is again applicable.

Production activity control

The work described in this paper is part of an ESPRIT (European Strategic Programme for Research in Information Technology) project titled 'Control Systems for Integrated Manufacturing'. The overall objective of this project is to produce an application generator (AG) which will provide tailor-made PAC systems for CIM environments. Approaching this problem from the 'outside-in' seemed somewhat insurmountable. An 'inside-out' approach seems to provide a better chance of success. This entails building a small PAC system and adapting it for other industries until a comprehensive system is developed. The AG will then be coded around this system. This mini-PAC system is called the breadboard as it will be developed in the electronics industry by the Digital Equipment Corporation, Eire and transferred for testing and modification to the automotive/metal removal industries (Comau, Italy, and Renault, France).

Functionality

The breadboard represents the functional requirements of PAC for the manufacturing system (see Fig. 3) in the electronics industry. The following are the main functional blocks which have been identified for the breadboard:

- *Scheduler* – Schedules orders for manufacturing system after receiving the initial requirements from MRP or other order developing system.
- *Dispatcher* – Real-time decision-maker of the breadboard. Working with knowledge of the production system and the order schedule, the dispatcher allocates jobs to machines and directs the flow of material through the manufacturing system.
- *Mover* – Controls material handling system under the direction of the dispatcher.
- *Monitor* – Data collection and monitoring.

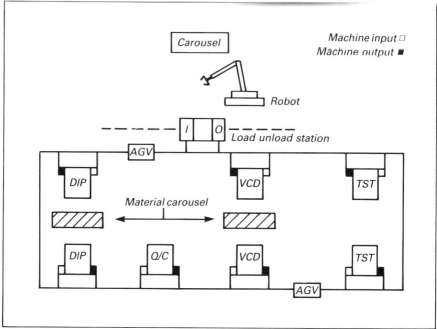

Fig. 3   Example manufacturing system

- *Manufacturing system simulator* – Provides the functionality of the production system and its components in the initial stages of development. This simulates the flow of parts and their interaction with the following entities:

  – Robot.
  – Carousel.
  – AGVs (automated guided vehicles).
  – Machines.

- *User interface* – Provides graphical and textual information on crucial parameters of the system.

From the above description of the breadboard, simulation can be seen to have two possible roles:

- Simulating the processes to be controlled (only in the initial states of the breadboard development).
- Controlling and scheduling within PAC.

The first case is a conventional simulation involving delays, processors, etc. However, it is driven by the second case, which produces control commands in reaction to events in controlling and scheduling is incorporated in the dispatcher and scheduler functions, with some of the controlling function carried out in the mover and monitor. For ease of communication between the two cases of simulation both are implemented in OPS5.

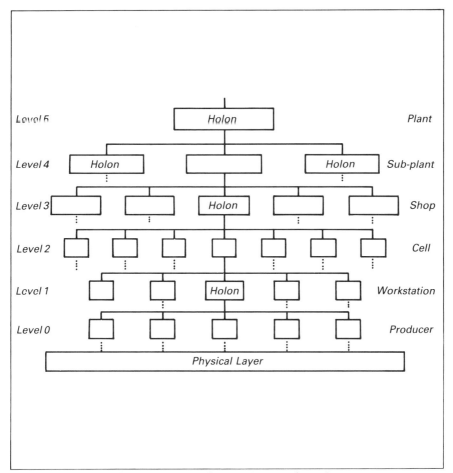

Fig. 4   Decision-making hierarchy of example manufacturing system

Control architecture

Basically PAC can be seen as the carrying out of the long-term plans from the master production schedule and materials requirement plan. It is a short-term activity and by nature highly detailed, or 'data-rich'. Because of the nature of the PAC problem a decentralised-hierarchical control architecture is utilised in this work (see Fig. 4). Control is exercised by and on the basis of communication between different levels and holons. (The term 'holon' represents a system which is itself a part of a higher system and is also a parent of other systems.) The functional blocks mentioned above are mapped onto this control architecture, with each holon containing aspects of all functions. Thus at the shop level for instance there exist both a scheduler and a dispatcher concerned with producing and controlling schedules for the cells in the next level. The structure of these functions is similar at all levels but the data on which they operate differ. The next section describes the structure and operation of both.

Simulator

Entities in the production system are modelled using the data structure described above. For instance, a machine is represented in the world model as follows:

```
;;
   (Literalize Machine              ;Machine
                Type                ;Type of Machine
                Name                ;Machine ID Number
                Input _ Location    ;Location of Input Buffer
                Output _ Location   ;Location of Output Buffer
                Input _ Buffer      ;Contents of Input Buffer
                Output _ Buffer     ;Contents of Output Buffer
                Status              ;Whether Idle or Busy
                Raw _ Materials     ;List of processable Raw Materials
                Fixture _ Equip     ;List of required Fixture Equipment
                Operations)         ;List of possible Operations
;;
```

In this example the entity name is 'Machine' with attributes 'Type' through to 'Operations'. An entity is created by assigning a value to any one attribute, and its attribute values can be modified during the simulation.

Simulation events are generated from two sources. First, the control functions pass events to the simulated processes, such as:

– Move AGV to Location A27,
– Remove Part from Carousel.
      :
      .

Other events are created within the manufacturing system simulation itself, such as:

– End of a Machine Operation,
– Random events, e.g. Machine Breakdown.
      :
      .

It is helpful to implement both simulation in OPS5 so as to facilitate event and message passing and to avoid duplication of data. The manufacturing system simulation is discrete-event driven where the events are data structures as below:

```
(Event        ^Type      Create _ Job
              ^Time      6)
```

This is similar in manner to conventional simulation languages, such as SLAM[12], where events have event codes and event times. Rules to simulate changes in the state of the world model are fired when the following conditions are true:

- An event of a correct type exists and its activation time is the same as the simulation clock.
- All the left-hand-side conditions of the rule are true.

The right-hand-side actions of the rule modify the attribute values of and/or remove and create entities. This changes the state of the world model and advances the simulation. The following is an example of a rule to insert a part into the carousel:

```
;;

( P Job _ Inserted _ in _ Carousel          ;;;;;; IF
    (Clock^Time <Now>)                       ;Simulation Time
    (Event^Type Insert _ Job                 ;Event Type
            ^Time <Now>)                      ;Activation Time
    (Feeder^Type Robot                       ;Feeder – Robot
            ^Status Withdrawing              ;Present Task
            ^Content <Job>)                  ;Content of feeder
    (Carousel                                ;Carousel (Max = 8)
            ^Jobs _ Present { <Amount> < 8}) ;Check if < to 8
    (Carousel  _ Locations                   ;Space in carousel
            ^Location <Number>               ;with no Job
            ^Job = 0)
                                             ;;;;;; THEN
-->

(Remove 2)                                   ;Delete Event
(Modify 3^Content Empty                      ;Modify Robot
        ^Status Inserting)                   ;New Task
(Modify 4^jobs _ Present                     ;Update no. jobs
        (Compute (<Amount> + 1)))            ;in Carousel
(Modify 5^Job <Job>)                         ;Update Location
(Make Event^Type Finish _ Insert             ;Create Event to
        ^Time (Compute (<Now> + 1))))        ;simulate finish
                                             ;of insertion after
;;                                           ;1 time unit
```

The timing is achieved as in conventional simulation languages with a clock being incremeted to the value of the time attribute of the next event.

Decision-making

The dispatcher makes real-time decisions dependent on the current state of the world model. These decisions are made on the basis of simulation, which in contrast to the manufacturing system simulation is message or transaction driven. Transactions are returned from the manufacturing system requesting a decision. Based on the applicable strategy implemented as heuristics, events are created and passed to the manufacturing system. The following is an example of a rule to pass an event to unload a part from the carousel and transport it to the machine.

Based on the current state of the world model, the dispatcher may project the schedule forward a finite distance in time and thus simulate an action. The new state may be examined and, depending on its effectiveness, a decision can be made to backtrack or not. If the dispatcher decides to backtrack a different rule from the conflict resolution set will be fired. Once a state is achieved that is effective the dispatcher issues an event to the production system:

```
;;
(P   Dispatch _ 6                            ;;;;; IF
     (Transaction^Type Machine _ Idle)       ;Machine returned IDLE
     (Carousel^Jobs _ Present = 8)           ;Carousel is Full
     (Job^Number <Job _ Number>              ;Select Job (N.B.No
          ^Status Stored)                    ;Selection Strategy
     (Machine
               ^Name <Name>                  ;Machine entity which
               ^Input _ Buffer Empty         ;generated Transaction
               ^Status Inactive)
     (load/unload                            ;L/U Station Available
               ^Output _ Buffer Empty)
     (Feeder^Type AGV                        ;Feeder AGV Available
          ^Name AGV _ 1                      ;and in right location,
          ^Location Down _ out               ;hence no need to move
          ^Content Empty                     ;and it is empty
          ^Status Inactive)

-->                                          ;;;;; THEN
     (Modify 3^Status Removing)              ;Status of Job is changed
     (Modify 4^Status Assigned)              ;Machine is assigned
     (Make  Event^Type Remove _ Job          ;Create Event to remove
          ^Time (Compute <Now> + 1)))        ;job from Carousel

     ;;
```

From the scheduling point of view, orders are scheduled according to different scheduling strategies. These various schedules may be simulated to determine the best selection for the current state of the world model. Abilities exist in OPS5 (Savestate, Restorestate) to save and later restore the current state of the world model, and this makes simulations of this type feasible. The scheduler may be used in a 'what-if' manner to test the reaction of the manufacturing system to variations in load and capacity.

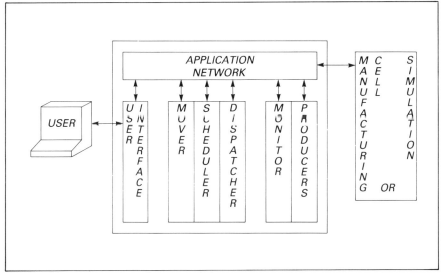

Fig. 5 Abstract computational model

Future developments

The functional blocks described above are mapped onto software modules to give the abstract computational model of Fig. 5. These elements are independent entities each with a specific responsibility which communicate by generating and receiving transactions. It is the responsibility of the application network to ensure that each transaction is delivered to the entity or entities which it impacts.

Given the modular nature of the breadboard it will be relatively easy to replace the simulator with representations of different manufacturing facilities. At present the possibility of utilising petri nets to describe the facility and then translating the petri net into a simulation model is been examined.

It is proposed to build an AI environment around the breadboard. The main objective of which will be to simulate differing scheduling and/or dispatching strategies and tactics so as to provide quantitative and qualitative measures of performance. These measures will aid in the decisions as to which strategies are most applicable.

Concluding remarks

The all-important distinction between the simulations described here and conventional simulation lies in their emphasis on information flow rather than entity flow. The variable and uncertain nature of this information flow in PAC makes it a suitable application space for AI.

With regard to conventional simulation languages, the following are some of their main differences. Conventional simulation supports:

- Primarily numeric processes.
- Algorithmic solutions.

- Integration of data and control structure.
- Simulation of part and process flow.

AI-based simulation on the other hand facilitates:

- Primarily symbolic processes.
- Non-explicit solutions.
- Separation of knowledge and control structure.
- Simulation of information flow.

In conclusion, the usefulness of AI-based simulation of manufacturing systems has been demonstrated and particularly two roles within a computer-integrated manufacturing environment have been identified:

- Scheduling in automated systems.
- Real-time on-line control system.

Similarly, the nature of the decentralised, hierarchical control architecture in PAC, where information flow between the levels and holons is important, suggests the need for AI due to its ability to model the flow of information. If the attention devoted to AI in recent years is successful in achieving the goals AI has set itself, these changes will have an important impact on simulation.

Acknowledgements

This work is funded by the Digital Equipment Corporation BV (Eire) and the European Economic Community funded ESPRIT programme.

References

[1] Shannon, R. and Mayer, R. 1985. Expert systems and simulation. *Simulation* June 1985: 275-284.
[2] Gaines, B. and Shaw, M. 1985. Expert systems and simulation. In, *AI, Graphics and Simulation*, 1985.
[3] Baskaran, V. and Reddy, Y. 1984. An introspective environment for knowledge based simulation. In, *Winter Simulation Conf.*, 1984.
[4] Vaucher, J. 1985. Views of modelling: Comparing the simulation and AI approaches. In, *AI, Graphics and Simulation*, 1985.
[5] Adelsberger, H. 1984. Prolog as a simulation language. In, *Winter Simulation Conf.*, 1984.
[6] Radzikowski, P. 1984. Framework of the decision support expert systems. In, *Winter Simulation Conf.*, 1984.
[7] Elmaghraby, A. and Jagannathan, 1985. An expert system for simulationists. In, *AI, Graphics and Simulation*, 1985.
[8] Cleary, J., Goh, K. and Unger, B. 1985. Discrete event simulation in Prolog. In, *AI, Graphics and Simulation*, 1985.
[9] Harhen, J. and Browne, J. 1983. Production Activity Control – A key node in CIM (Computer Integrated Manufacturing). In, *Int. Working Conf. on Strategies for Design and Economic Analysis of Computer-Supported-Production-Management Systems (PMS)*, Vienna, 28–30 September 1983.

[10] Harhen, J. 1983. Production Activity Control and the new way of life. *Production & Inventory Management*, 24: 73-85.

[11] Keen, P. and Morton, M. 1978. *Descision Support System: An Organised Perspective*. Addison-Wesley, Reading, MA, USA.

[12] Pritsker, A.A.B. 1984. *Introduction to Simulation and SLAMII*. 2nd edn. Halstead Press, New York.

EXPERT SYSTEMS AND SIMULATION IN THE DESIGN OF AN FMS ADVISORY SYSTEM

B.R. Gaines
University of Calgary, Canada

A system combining expert system, database and simulation techniques is described which acts as a decision support system for flexible manufacturing systems. The system interacts with the manager of a computer integrated manufacturing system both at the equipment procurement stage and at the system operation stage to aid him in planning machine acquisition, placement and operation. It consists of four modules: a database of machinery; a machine grouping optimiser; a simulator of flexible manufacturing system configurations; and an expert system shell containing advisory rules on flexible manufacturing systems. This paper gives the design considerations for the system, describing the expert system architecture, presenting the knowledge engineering considerations involved in its development, and illustrating its operation.

Expert systems (ESs) have become widely accepted as a significant new computing technology with high commercial potential and important industrial applications[45,57]. However, there is a paradox in current ESs, for in their greatest strength lies their greatest weakness. The encoding of the superficial, 'what to do' knowledge underlying the performance of a skill, rather than the deep, 'why to do it' knowledge underlying the generation of optimal algorithms, has allowed ES technology to be applied to a variety of problems that were previously intractable. However, 'what to do' knowledge is subject to far more rapid obsolescence with change than is 'why to do it' knowledge, and the problem of keeping ESs up-to-date may more than offset their advantages in many applications. There are also many problems where neither the superficial nor the deep approaches alone are adequate.

The most difficult problem for the computing industry currently is to determine what classes of applications are amenable ESs. Similar problems do not necessarily have similar solutions. One application may be eminently suited to the ES approach and another that differs apparently in only minor details may be intractable. Our conceptual models of the application domain give it a topology that is full of holes. We need reconceptualise our approach to the use of information technology in decision and control problems.

Simulation technology has been based largely on encoding the deep mathematical models underlying physical processes and the extension of ESs to incorporate simulation is one step towards stengthening ESs. However, to create hybrid ES/simulation systems alone will leave holes in the applications domain that are still non-obvious. We need deeper foundations that give us both powerful and comprehensible technologies. This paper considers the development of systems integrating ESs, simulation and other information technologies.

The ES paradigm shift

The change in approach represented by ES can be seen in the context of the classical decision and control paradigm: first model the system then optimise decision and control against that model. The ES breakthrough was to realise that in certain problems where the controlled system cannot be modelled but people can do the task, it is possible to automate by modelling the person – that is, the new paradigm is to model the controller not the plant[15,16].

This new paradigm has proved very powerful in enabling a number of effective decision and control systems to be developed for problems that were previously intractable[14,15]. The range of ES applications is now very wide and it is possible to see some patterns in successful systems. Hayes-Roth typifies successful ES applications as problems with the following characteristics[23]:

● The organisation requires more skilled people than it can recruit or retain.
● Problems arise that require almost innumerable possibilities to be considered.
● Job excellence requires a scope of knowledge exceeding reasonable demands on human training and continuing education.
● Problem solving requires several people because no single person has the needed expertise.
● The company's inability to apply its existing knowledge effectively now causes management to work around basic problems.

All these characteristics presume that:

● The knowledge required is available to the organisation.
● The problem is ineffective use of the knowledge through lack of people or lack of coordination.
● The knowledge can be made overt (knowledge engineering).
● The knowledge can be encoded for use in an ES (power of ES shell).

The strength of ES applications has been in the large number of situations that satisfy these requirements. The weakness is the limitations of current ES technology that make the last one, knowledge representation, a major impediment to many applications. These limitations stem from the very paradigm shift that was noted above as the breakthrough created by ESs, modelling the controller or decision-maker rather than the plant, or system about which decisions are being taken. Current ES shells can only operate within this paradigm and hence they do not have the capability to incorporate the knowledge that was the basis of most classical decision and control systems, the physical models of the relevant worlds.

In the ES literature this has been termed the problem of *deep knowledge*[8], of going beneath the pragmatic rules used by people to carry out a skilled task and incorporating the more foundational premises on which those rules are based. The problem shows up in systems like MYCIN[6] for diagnosing microbial infections, where the rules pre-suppose fixed universes of infections, diagnostic techniques and drugs. If any of these universes changes then the system has no way to respond. It is not just a problem of adding some additional rules. The existing rules may need to be edited because they pre-suppose that certain conditions do not exist. The entire rule base may have to be developed again as a result of apparently minor changes.

Superficial knowledge may be adequate to represent an existing skill but inadequate to encode the capability to adapt to change. Individual people may show similar characteristics in coping with change. However, society as a whole has the capability to adjust to new conditions and disseminate the knowledge required to cope with change. ESs could be incorporated in this process by posting updates to their rule bases through a communications network, and this will happen for some large-scale applications. However, the task of keeping very large numbers of highly specialised and customised ESs up-to-date by editing rule bases at the current level of superficiality is impossible. The developmental pressure is towards deeper knowledge and the integration of ESs with existing information technologies, particularly numeric computation, databases and simulation[45].

This paper examines the integration of expert systems technology with other information processing technologies, particularly simulation, in the context of an advisory system concerned with the design and management of computer-integrated manufacturing systems. In particular, the knowledge engineering involved in the overall system design is described.

Computer-integrated manufacturing

We are in the midst of an intellectual revolution in manufacturing. Developments in numerically controlled machine tools, material handling, industrial robotics, and the understanding of human aspects of productivity and quality assurance, have led to new approaches to manufacturing, the cumulative effect of which is revolutionary rather than evolutionary[2,22,25,29,44,48,56].

A number of major changes in manufacturing technology are apparent:

• There has been a shift from mass-production transfer lines highly optimised for long runs of a particular product towards flexible manufacturing units coping cost-effectively with shorter runs of a range of products[9].

• There has been a shift from discrete manufacturing processes towards continuous flow processes based on robotic material handling[11,28].

• There has been a shift from production of piece parts, perhaps at different specialist sites with subassembly and assembly at other sites, towards single-site integrated manufacturing[1,34].

• There has been a shift from high-volume, single-site manufacturing and widespread geographic distribution towards lower-volume manufacturing replicated at multiple sites chosen to minimise distribution costs[34,52].

• There has been a shift from the use of people as attachments to production lines and doing small repetitive sub-tasks towards the use of people in a supervisory role with emphasis on responsibility for complete products[34,47].

Computer-integrated manufacturing (CIM) support such shifts by allowing automation and mass-production techniques to be applied to fairly short runs of diverse items compared with the long runs of limited variety required in the past[23]. The flexibility, customisation and localisation offered are attractive but generate a new class of management problems[54]. The information flows required in manufacturing increase greatly with the diversity of product being handled and the rapidity of change between products. The knowledge required to manage these information flows is not readily available. Setting up a flexible production line requires different and higher-level skills than those required to operate it once running. A variety of additional skills are required to design a flexible manufacturing system capable of handling a diversity of products[24,35,42].

Management support systems are needed for the planning, implementation, configuring and operation of CIM[10,41,43]. Expert systems provide a technology to cope with the diversity and complexity of the information involved[31,36,39,40]. However, an expert system alone cannot cope with the volume of information about different machine tools and robots, the computation required in grouping and the provision of numeric simulations for interactive planning and testing. An adequate support system for managing CIM involves the combination of expert system and simulation-based decision aids with numeric computation, databases, communications, and, if it is to be used in operations, instrumentation and control systems[37,46,53].

Expert system technology can be applied at a number of levels in CIM including product design[32], flexible manufacturing system (FMS) justification[55], job-shop scheduling[12], simulation[5], on-line process control[4] and robot control[33]. However, the problem of major concern is the design stage at which the manufacturing system has to be specified in terms of its performance and a range of diverse machine tools have to be brought together in an integrated system design. This stage presents a number of major conceptual difficulties that are currently impeding the rapid diffusion of new manufacturing technology into appropriate industries. It is typical of the problems that because of their complexity and diffuseness they have to be 'managed' and are very susceptible to support through expert systems[26].

The following section describes the development of advisory systems combining expert system, database and simulation technologies in order to support the decision processes of managers responsible for configuring FMSs.

A technology management system for FMS planning

The considerations above have led to the design of a decision support system for planning FMS which is part of a larger project on the development of *Technology Management Systems*. The system combines database, optimisation, simulation, knowledge base, expert system and human-computer interaction technologies in the decision-support system shown in Fig. 1.

The Technology Management System is designed to:

- Provide a simple, straightforward human-computer interface to a variety of tools for the design and management of an FMS.
- Use a knowledge-base of FMS design and procurement procedures to elicit the critical factors in the requirements for an FMS such as throughput, workpiece sizes, machining activities, precision requirements, and so on.
- Use an expert system to advise on the feasibility of the applications, possible configurations and the types of equipment required.
- Use a database of machine tools and robots to match the requirements range against a variety of tool and material handling configurations.
- Compute the optimum grouping of tools to meet the specified requirements.
- Make available through simulation, operating parameters such as set-up times, throughput rates, tolerance range, failure rates, and so on, for a variety of jobs within the requirements range.
- When reasonable matches between requirements and performance are found, refine the design at a greater level of detail including operational factors and overall cost-effectiveness for a variety of job mixes and production runs.
- Finally, produce full specifications of the FMS including set-up procedures.

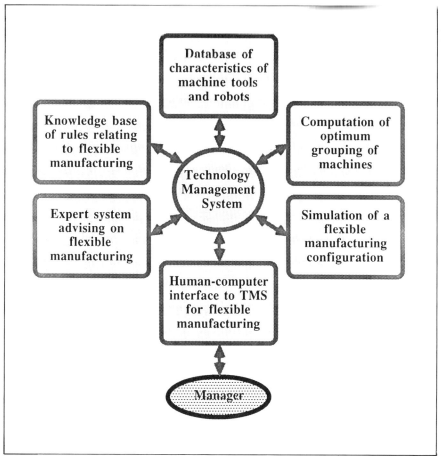

Fig. 1   Technology Management System for planning flexible manufacturing systems

Knowledge engineering for an FMS Technology Management System

The systems analysis required for a system encompassing the specified capabilities is a major task. This type of analysis has come to be termed *knowledge engineering* in the expert systems literature because it involves eliciting from experts the detailed knowledge processes that they use in performing the tasks involved[17,18,19,27,50,51]. What is characteristic of major real-world systems, however, is that the knowledge required for their operation is not localised within a single expert but distributed over a number of specialists with diverse skills in different parts of the organisation. In expert system terms knowledge engineering is required over a very wide domain which cannot be encompassed by one person.

To cope with this problem it is necessary to fragment the problem domain into a number of *knowledge islands* which are characterised by their internal coherence – within an organisation they usually correspond to functional specialities. The knowledge engineering for each island generally involves interviews with only one person or, if the relevant material is well-codified, analysis of a particular handbook or manual – the *Flexible Manufacturing Systems Handbook*[9] is a prime example of such codified knowledge. We have implemented techniques for carrying out the interviews automatically through computer interaction with the experts that elicit their constructs, their basic dimensions of thinking[13,14,50,51], and are developing techniques for deriving the construct systems embodied in text automatically. The implicational structure of the construct systems can also be derived automatically and used to generate rules priming an expert system shell[3,20].

Knowledge islands for FMS

Fig. 2 shows a preliminary analysis of the main knowledge islands involved in a technology management system for planning FMS.

The *overall requirements specification* is the starting point of any feasibility study of FMS. The key constructs here were found to be the

Fig. 2   Knowledge islands and areas underlying the knowledge engineering of expert systems supporting the planning of flexible manufacturing systems

obvious pairs *low-volume – high-volume* and *low-variety – high-variety*, but other significant constructs were also elicted that affect the feasibility of FMS implementation such as *FMS-already-in-use – FMS-not-in-use* and *committment-to-existing-machines – new-plant*. The experience dimension and the difference between restructuring and green-field situations are obviously relevant to the decision-making but could be missed in an analysis that concentrated only on quantitative parameters.

The experience/resources dimension refines naturally to the knowledge island concerned with *overall resources and constraints*. The volume/variety dimension refines to a quantitative description of the *parts specification* in parts per year and the number of different parts. Fig. 3 shows the conventional wisdom about the suitability of different forms of automated manufacturing dependent on the parts specification. Coding the knowledge involved for a computer shows up several

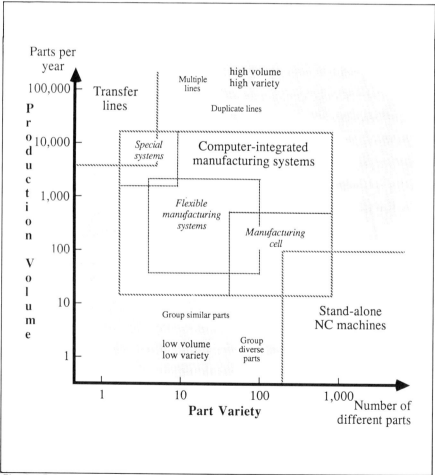

Fig. 3   *Primary variables affecting choice of manufacturing systems and strategies when outside ranges*

features that tend to be forgotten in the conventional analysis. For example, what advice do you give when a requirement falls outside all the boxes? – the answer is to restructure the requirement by grouping or splitting as shown by the grey areas in Fig. 3. Another complication is that part variety is not adequately represented by number of different parts – similarities between parts are very significant in system organisation and feasibility so boundaries based on numbers alone are fuzzy. They can be modified by qualitative considerations of part groupings well before they are refined to detailed considerations of machining requirements.

Consideration of the *resources* and *parts* islands can give enough information to prime the *overall system architecture* island based on a more detailed version of Fig. 3. This in its turn can lead to *rough performance estimates*. Surprisingly it can also provide sufficient data for *rough simulation* using the fuzzy linguistic model approach[16,58]. The *performance estimates* and *simulation presentation* form a basis for conclusions about the feasibility of an FMS for a particular requirement and the six knowledge islands involved may be grouped together as a *feasibility study* area. This subsystem is already a useful tool for the preliminary stages of FMS planning.

The knowledge islands described may be refined to a further level of detail on two main paths. The macro description of resources and constraints may be refined to a micro description that lists floor plans, personnel, machines, and so on. Much of this will be available in existing administrative databases but some information will probably be missing. The macro description of parts may be refined to a micro description that completely characterises each part and its manufacturing process[7]. Much of this will be available in existing corporate engineering and production databases but again some will be missing.

The *parts specification* leads naturally to a *manufacturing techniques* knowledge island which itself leads to that for tools required[30]. The parts and tools islands form another useful subsystem, the grouping area, where the optimal tool grouping knowledge island takes the detailed specification of parts and their machine utilisation requirements and uses these to derive nearly optimal groupings of machinery.

The central region of detailed architecture, performance and simulation is complicated more by the data requirements than by the complexity of computation required. It is unlikely to be cost-effective for any organisation to gather the detail required on a wide range of machines and field-experience (although as widespread use of the manufacturing automation protocol, MAP[21,38], develops much more data is potentially available economically). The knowledge islands in this area will evolve through successive refinements of those already described in a way determined by mainstream industry trends and cost-effectiveness. The phenomenon where leading suppliers tend to be used again because there is more knowledge of their products is one phenomenon of this region.

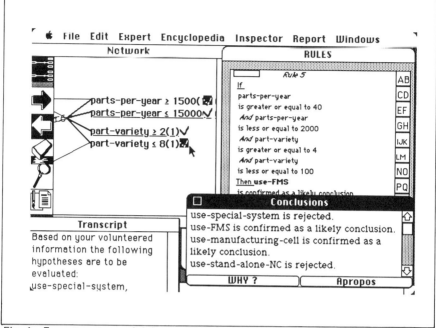

Fig. 4   *Expert system advising on FMS feasibility*

Implementation

The *feasibility study area* is being implemented on an Apple Macintosh in the *Nexpert* expert system shell which offers highly interactive facilities for forward and backward inference, hypothesis formation, explanation and graphical presentation of the structure of inferences and the knowledge base. Fig. 4 shows a Macintosh running a consultation in Nexpert on the overall requirements specification and system architecture. The database of machine tools and parts specifications will require facilities that go beyond those of either of these systems[41] and is being implemented on an Appollo network. The FMS simulation is being written in SIMSCRIPT II.5[59] as a module which takes in the conclusions of the expert system and gives a dynamic graphic presentation of the FMS operation and its performance parameters. The *grouping area* subsystem is being implemented using algorithms previously described[35] on an Appollo workstation which offers good facilities for simulation, graphic interaction and rule-based expert system specification.

Concluding remarks

Developments in computers, robotics, communications, electronics and materials are occurring at an increasing pace and have widespread impact on industrial manufacturing processes. In particular the combination of these new technologies offers opportunities for integrated manufacturing systems that increase productivity through improved reliability, reproducibility, flexibility, human factors and cost-effectiveness. CIM is applicable to a spectrum of industry from comparatively small firms up to large organisations. However, it is difficult for firms to keep up with the flow of information and experience on these rapidly changing new technologies.

Decision support systems for CIM present particular problems for today's expert system technology because manufacturing involves the dynamics of physical processes and requires knowledge based systems that combine simulation techniques with those of current expert systems. They also present challenges in coordination because the expertise required is dispersed across a number of disciplines, between universities and industry, and between equipment manufacturers and users. However, the relevant technologies and knowledge are now at a stage where practical systems may be developed. The refinements of these systems through experience in their use, further analysis and data collection, will lead to the evolution of the totally-integrated manufacturing system.

Acknowledgements

Financial assistance for this work has been made available by the National Sciences and Engineering Research Council of Canada. The author is grateful to his colleagues, Drs Kolodny, Moray, Shaw, Turksen and Vanelli for discussion and comments.

References

[1] Aguren, S. and Edgren, J. 1980. *New Factories*. Swedish Employers' Confederation, Stockholm.
[2] Beni, G. and Hackwood, S. (Eds.) 1985. *Recent Advances in Robotics*. John Wiley, New York.
[3] Boose, J.H. 1984. Personal construct theory and the transfer of human expertise. In, *Proc. AAAI-84*, pp. 27-33. American Association for Artificial Intelligence, CA, USA.
[4] Brown, I., Alexander, S.M., Jagannathan, V. and Kirchner, R. 1985. Demonstration of an expert system for manufacturing process control. In, Birtwistle, B. (Ed.), *AI, Graphics and Simulation*, pp. 110-113. Society for Simulation Research, La Jolla, CA, USA.
[5] Bruno, G., Elia, A. and Laface, P. 1985. A rule-based system for production scheduling. In, *Proc. COMPINT'85, IEEE*, pp. 35-39.
[6] Buchanan, B.G. and Shortliffe, E.H. (Eds.) 1984. *Rule-Based Expert Systems. The MYCIN Experiments of the Stanford Heuristic Programming Project*. Addison-Wesley, Reading, MA, USA.
[7] Carringer, R.A. 1985. Integrating manufacturing systems using the product definition interface. In, *Autofact'85*, pp. 1-17. SME, Dearborn, MI, USA.
[8] Chandrasekaran, S. and Mittal, S. 1984. Deep versus compiled knowledge approaches to diagnostic problem-solving. In, Coombs, M.J. (Ed.), *Developments in Expert Systems*, pp. 23-34. Academic Press, London.

[9] *Flexible Manufacturing Systems Handbook*. Noyes Publications, Park Ridge, NJ, USA, 1984.

[10] Elavia, J.D. 1985. Where do I start with CIM? In, *Proc. Autofact'85*, pp. 1.1-1.21. SME, Dearborn, MI, USA.

[11] Engelberger, J.F. 1980. *Robotics in Practice: Management and Applications of Industrial Robots*. Amacom, New York.

[12] Fox, M.S. 1985. Knowledge representation for decision support. In, Methlie, L.B. and Sprague, R.H. (Eds.), *Knowledge Representation for Decision Support Systems*, pp. 3-26. North-Holland, Amsterdam.

[13] Gaines, B.R. and Shaw, M.L.G. 1981. New directions in the analysis and interactive elicitation of personal construct systems. In, Shaw, M.L.G. (Ed.), *Recent Advances in Personal Construct Technology*, pp. 147-182. Academic Press, London.

[14] Gaines, B.R. and Shaw, M.L.G. 1984. Logical foundations of expert systems. In, *Proc. IEEE Conf. on Systems, Man and Cybernetics*, pp. 238-247.

[15] Gaines, B.R. and Shaw, M.L.G. 1985. Expert systems and simulation. In, Birtwistle, G. (Ed.), *AI, Graphics and Simulation*, pp. 95-101. Society for Simulation Research, La Jolla, CA, USA.

[16] Gaines, B.R. and Shaw, M.L.G. 1985. From fuzzy sets to expert systems. *Information Science*, 36(1-2): 5-16.

[17] Gaines, B.R. and Shaw, M.L.G. 1985. Foundations of expert systems. In, Kacprzyk, J. and Yager, R.R. (Eds.), *Management Decision Support Systems Using Fuzzy Sets and Possibility Theory*, pp. 3-24. ISR Interdisciplinary Systems Research, Verlag TUV.

[18] Gaines, B.R. and Shaw, M.L.G. 1985. Foundations of logic and reasoning in knowledge-based systems. In, Bezdek, J. (Ed.), *Proc. Int. Conf. on Fuzzy Information Processing*. To appear. CRC Press.

[19] Gaines, B.R. and Shaw, M.L.G. 1985. Systemic foundations for reasoning in expert systems. In, Gupta, M.M., Kandel, A., Bandler, W. and Kiszka, J.B. (Eds.), *Approximate Reasoning in Expert Systems*, pp. 271-281. North-Holland, Amsterdam.

[20] Gaines, B.R. and Shaw, M.L.G. 1986. Induction of inference rules for expert systems. *Fuzzy Sets and Systems*, 18(3): 315-328.

[21] General Motors, 1984. *General Motors Manufacturing Automation Protocol*. General Motors Corp., Warren, MI, USA.

[22] Gomersall, A. and Farmer, P. 1984. *Robotics: The International Bibliography with Abstracts*. IFS (Publications) Ltd, Bedford, UK.

[23] Groover, M.P. and Zimmers, E.W. 1984. *CAD/CAM: Computer-Aided Design and Manufacturing*. Prentice-Hall, Englewood Cliffs, NJ, USA.

[24] Gunn, T.G. 1981. *Computer Applications in Manufacturing*. Industrial Press, New York.

[25] Hartley, J. 1984. *Flexible Automation in Japan*. IFS (Publications) Ltd, Bedford, UK.

[26] Hayes-Roth, F. 1984. The industrialization of knowledge engineering. In, Reitman, W. (Ed.), *Artificial Intelligence Applications for Business*, pp. 159-177. Ablex, Norwood, NJ, USA.

[27] Hayes-Roth, F., Waterman, D.A. and Lenat, D.B. (Eds.) 1983. *Building Expert Systems*. Addison-Wesley, Reading, MA, USA.

[28] Heginbotham, W.B. (Ed.) 1984. *Programmable Assembly*. IFS (Publications) Ltd, Bedford, UK.

[29] Kafrissen, E. and Stephans, M. (Eds.) 1984. *Industrial Robots and Robotics*. Reston, VA, USA.

[30] Kawabe, S., Okano, A. and Yoshida, T. 1985. Robot task planning system based on product modelling. In, *Proc. COMPINT'85, IEEE*, pp. 471-476.

[31] Kempf, K.G. 1985. Manufacturing and artificial intelligence. *Robotics*, 1(1): 13-25.

[32] Kim, S.H. and Suh, N.P. 1985. On an expert system for design and manufacturing. In, *Proc. COMPINT'85, IEEE*, pp. 89-95.

[33] Kirschbrown, R.H. and Dorf, R.C. 1985. KARMA – a knowledge-based robot manipulation system. *Robotics*. 1(1): 3-12.

[34] Kolodny, H.F. 1985. Three bases for QWL improvements. University of Toronto, Faculty of Management Studies.

[35] Kusiak, A., Vannelli, A. and Kumar, K.R. 1985. Grouping problems in scheduling flexible manufacturing systems, *Working Paper 84-16*. University of Toronto, Department of Industrial Engineering.

[36] Liu, D. 1985. Intelligent manufacturing planning systems. In, *Autofact'85*, pp. 1-17. SME, Dearborn, MI, USA.

[37] Maimon, O. 1985. FMS real time operational control. In, *Proc. Autofact'85*, pp. 6.31-6.58. SME, Dearborn, MI, USA.

[38] McCarthy, J.J. 1985. MAP's impact on process plants. *Control Engineering*, 32(11): 67-69.

[39] McDonald, M.E. 1985. The smart foreman – expert systems for manufacturing. In, *Proc. Autofact'85*, pp. 1-9. SME, Dearborn, MI, USA.

[40] Miller, R.D. 1985. Simplification, management control, then intelligent automation. In, *Proc. Autofact'85*, pp. 1-14. SME, Dearborn, MI, USA.

[41] Mills, R.B. 1985. Database machines tackle the CAD/CAM glut. *Computer-Aided Engineering*, 4(10): 61-64.

[42] Morgan, C. 1984. *Robots: Planning and Implementation*. IFS (Publications) Ltd, Bedford, UK.

[43] Nassr, J.J. 1985. Considerations in establishing a CIM environment. In, *Proc. COMPINT'85, IEEE*, pp. 68-72.

[44] Rathmill, K., MaConail, P., O'Leary, S. and Browne, J. (Eds.) 1985. *Robotic Technology and Applications*. Springer-Verlag, Berlin.

[45] Reitman, W. (Ed.) 1984. *Artificial Intelligence Applications for Business*. Ablex, Norwood, NJ, USA.

[46] Rhodes, J.S. 1985. The development of integrated control systems for flexible manufacturing systems. In, *Proc. Autofact'85*, pp. 6.11-6.29. SME, Dearborn, MI, USA.

[47] Rubenowitz, S., Norrgren, F. and Tannenbaum, A.S. Some social psychological effects of direct and indirect participation in ten Swedish companies. *Organization Studies*, 4(3): 243-259.

[48] Scott, P.B. 1984. *The Robotics Revolution*. Basil Blackwell, Oxford, UK.

[49] Shannon, R.E., Mayer, R. and Adelsberger, H.H. 1985. Expert systems and simulation. *Simulation*, 44(6): 275-284.

[50] Shaw, M.L.G. and Gaines, B.R. 1986. Interactive elicitation of knowledge from experts. *Future Computing Systems*, 1(2), to appear.

[51] Shaw, M.L.G. and Gaines, B.R. 1986. An interactive knowledge elicitation technique using personal construct technology. In, Kidd, A. (Ed.), *Knowledge Elicitation for Expert Systems: A Practical Handbook*. To appear. Plenum Press, New York.

[52] Skinner, W. 1974. The focussed factory. *Harvard Business Review*, May: 113-121.

[53] Spur, G. and Mertins, K. 1985. Interactive production control in CAM. In, *Proc. COMPINT'85, IEEE*, pp. 572-576.

[54] Sullivan, W.A. 1985. Computer integrated manufacturing program justification. In, *Proc. Autofact'85*, pp. 2.34-2.55. SME, Dearborn, MI, USA.

[55] Sullivan, W.G. and LeClair, S.R. 1985. Justification of flexible manufacturing systems using expert system technology. In, *Proc. Autofact'85*, pp. 7.17-7.13. SME, Dearborn, MI, USA.

[56] Tver, D.F. and Bolz, R.W. 1983. *Robotics Sourcebook and Dictionary*. Industrial Press, New York.

[57] Waterman, D.A. 1986. *A Guide to Expert Systems*. Addison-Wesley, Reading, MA, USA.

[58] Wenstop, F. 1981. Deductive verbal models of organizations. In, Mamdani, E.H. and Gaines, B.R. (Eds.), *Fuzzy Reasoning and its Applications*, pp. 149-167. Academic Press, London.

[59] *SIMSCRIPT II.5 Programming Language*. CACI, Los Angeles, 1983.

Authors' organisations and addresses

L. Anderen
Cromemco
Cambridge House
180 Upper Richmond Road
London SW15 2SH
England

R. Beadle
Istel Ltd
Highfield House
Headless Cross Drive
Headless Cross
Redditch B97 SEU
England

D. Ben-Arieh
AT&T Bell Laboratories
6200 East Broad Street
Columbus, OH 43213
USA

M.J. Birch
School of Computing
Lancashire Polytechnic
Preston PR1 2TQ
England

A.S. Currie
Department of Product
  Management and Manufacturing
  Technology
University of Strathclyde
James Weir Building
75 Montrose Street
Glasgow G1 1XJ
Scotland

T.S. Chan
Department of Production
  Management and Manufacturing
  Technology
University of Strathclyde
James Weir Building
75 Montrose Street
Glasgow G1 1XJ
Scotland

B.J. Clarke
Department of Mechanical and
  Production Engineering
Huddersfield Polytechnic
Queensgate
Huddersfield
West Yorkshire MD1 3DH
England

P.L.C. Dunn
Process and Automation Division
GEC Mechanical Handling Ltd
Birch Walk
Fraser Road
Erith, Kent DA8 1QH
England

B.R. Gaines
Department of Computer Sciences
University of Calgary
Calgary
Alberta T2N 1N4
Canada

T.C. Goodhead
Department of Engineering
University of Warwick
Coventry CV4 7AL
England

T.M. Gough
TI Raleigh Ltd
Triumph Road
Nottingham NG7 2DD
England

P.W. Udo Graefe
Systems Laboratory
Division of Mechanical
  Engineering
National Research Council of
  Canada
Ottawa K1A 0R6
Canada

N.R. Greenwood
General Electric Industrial
  Automation, Europe
Shortlands
Hammersmith
London W6 8BX
England

R. Griffin
Inbucon Management
  Consultants Ltd
The Venture Centre
Warwick Science Park
Coventry CV4 7EZ
England

R.W. Hawkins
Ford of Europe Inc.
Eagle Way
Brentwood
Essex CM13 3BW
England

R. Hawthorn
School of Engineering
The Polytechnic
Wolverhampton
Wulfruna Street
Wolverhampton WV1 1SB
England

S.R. Hill
ICI plc
PO Box 11
The Heath
Runcorn
Cheshire WA7 4QE
England

R.D. Hurrion
School of Industrial and Business
  Studies
University of Warwick
Coventry CV4 7AL
England

J.E. Lenz
CMS Research Inc.
945 Bavarian Court
Oshkosh, WI 54901
USA

R.I. Mills
Ingersoll Engineers Inc.
Bourton Hall
Rugby CV23 9SD
England

R.J. Miner
9080 Dewberry Court
Indianapolis, IN 46260
USA

K.J. Musselman
Pritsker & Associates Inc.
1305 Cumberland Avenue
PO Box 2413
West Lafayette, IN 47906
USA

G. Seliger
Fraunhofer-Institut für
  Produktionsanlagen und
  Konstruktionstechnik (IPK)
Pascalstrasse 8-9
1000 Berlin 10
West Germany

J. Shivnan
Digital Equipment
  International BV
Clonmel
Co. Tipperary
Eire

H.-J. Warnecke
Fraunhofer-Institut für
  Produktionstechnik und
  Automatisierung (IPA)
Nobelstrasse 12
D-7000 Stuttgart 80
West Germany

J.F. Wilson
Ingersoll Engineers Inc.
Bourton Hall
Rugby CV23 9SD
England

Source of material

Visual interactive modelling
First published in European Journal of Operational Research, Vol. 23, 1986.
Reprinted courtesy of the author and North-Holland Publishing Company.

Animated CAD
First presented at the 1984 Systems International Conference, November 1984.
Reprinted courtesy of the author and Systems International.

Applying simulation to assembly line design
Not previously published.

Computer simulation – A feasibility and planning tool for FMS
First presented at the 2nd International Conference on Flexible Manufacturing Systems, 26-28 October 1983, London, UK.
Reprinted courtesy of the author and IFS (Conferences) Ltd.
Updated September 1986.

Simulation and reduction of risk in financial decision-making
First presented at the 1st International Conference on Simulation in Manufacturing, 5-7 March 1985, Stratford-upon-Avon, UK.
Reprinted courtesy of the author and IFS (Conferences) Ltd.

Risk avoidance through independent simulation
First presented at the 2nd International Conference on Simulation in Manufacturing, 24-26 June 1986, Chicago, USA.
Reprinted courtesy of the author and IFS (Conferences) Ltd.
Updated September 1986.

Simulation as a CIM planning tool
First presented at CIMCOM '85, 15-18 April 1985, Anaheim, California, USA.
Reprinted courtesy of the author and the Society of Manufacturing Engineers (SME-MS-85-332).

Simulation of a robotic welding cell for small-batch production

First presented at the 7th British Robot Association Annual Conference, 14-16 May 1984, Cambridge, UK.
Reprinted courtesy of the author and the British Robot Association.
Updated September 1986.

Modelling of a controller for a flexible manufacturing cell

Not previously published.

Manufacturing cell performance simulation using ECSL

Not previously published.

Simulation as an integral part of FMS planning

First presented at the 2nd International Conference on Simulation in Manufacturing, 24-26 June 1986, Chicago, USA.
Reprinted courtesy of the authors and IFS (Conferences) Ltd.

Decision support for planning flexible manufacturing systems

First presented at the 2nd International Conference on Simulation in Manufacturing, 24-26 June 1986, Chicago, USA.
Reprinted courtesy of the authors and IFS (Conferences) Ltd.
Updated September 1986.

'FORSSIGHT' and its application to an FMS simulation study

First presented at the 1st International Conference on Simulation in Manufacturing, 5-7 March 1985, Stratford-upon-Avon, UK.
Reprinted courtesy of the authors, British Aerospace plc and IFS (Conferences) Ltd.

Introducing FMS by simulation

First presented at the 2nd International Conference on Flexible Manufacturing Systems, 26-28 October 1983, London, UK.
Reprinted courtesy of the authors and IFS (Conferences) Ltd.

Experience in the use of computer simulation for FMS planning

First presented at the 1st International Conference on Simulation in Manufacturing, 5-7 March 1985, Stratford-upon-Avon, UK.
Reprinted courtesy of the authors and IFS (Conferences) Ltd.

Designing the control of automated factories

First presented at AUTOFACT 4, 4-7 November 1983, Detroit, USA.
Reprinted courtesy of the author and the Society of Manufacturing Engineers (SME-MS-83-789).

The use of simulation in cycle manufacture

First presented at the 1st International Conference on Simulation in Manufacturing, 5-7 March 1985, Stratford-upon-Avon, UK.
Reprinted courtesy of the author and IFS (Conferences) Ltd.

A production control aid for managers of manufacturing plants

First presented at the 1st International Conference on Simulation in
Manufacturing, 5-7 March 1985, Stratford-upon-Avon, UK.
Reprinted courtesy of the authors and IFS (Conferences) Ltd.
Updated September 1986.

Simulation on microcomputers – The development of a visual interactive modelling strategy

First presented at the Winter Simulation Conference '84.
Reprinted courtesy of the authors and the Board of Directors of the Winter
Simulation Conference.

Computer simulation for FMS

First presented at the 1st International Conference on Simulation in
Manufacturing, 5-7 March 1985, Stratford-upon-Avon, UK.
Reprinted courtesy of the authors and IFS (Conferences) Ltd.
Updated September 1986.

FMS: What happens when you don't simulate

First presented at the 1st International Conference on Simulation in
Manufacturing, 5-7 March 1985, Stratford-upon-Avon, UK.
Reprinted courtesy of the author and IFS (Conferences) Ltd.

Choosing and using a simulation system

First presented at the 3rd European Conference on Automated
Manufacturing, 14-16 May 1985, Birmingham, UK.
Reprinted courtesy of the authors and IFS (Conferences) Ltd.
Updated September 1986.

Practical experience contrasting conventional modelling and data-driven visual interactive simulation techniques

First presented at the 2nd International Conference on Simulation in
Manufacturing, 24-26 June 1986, Chicago, USA.
Reprinted courtesy of the authors and IFS (Conferences) Ltd.

A knowledge-based system for simulation and control of FMS

First presented at the 2nd International Conference on Simulation in
Manufacturing, 24-26 June 1986, Chicago, USA.
Reprinted courtesy of the author and IFS (Conferences) Ltd.

AI-based simulation of advanced manufacturing systems

First presented at the 2nd International Conference on Simulation in
Manufacturing, 24-26 June 1986, Chicago, USA.
Reprinted courtesy of the authors and IFS (Conferences) Ltd.
Updated September 1986.

Expert systems and simulation in the design of an FMS advisory system

First presented at the 2nd International Conference on Simulation in
Manufacturing, 24-26 June 1986, Chicago, USA.
Reprinted courtesy of the author and IFS (Conferences) Ltd.
Updated September 1986.